WHY DESIGN IS HARD

Why Design Is Hard

SCOTT BERKUN & BRYAN ZUG

A Schrödinger's Polyp Production
Berkun Media

Copyright © 2024 by Scott Berkun

All rights reserved. No part of this book may be reproduced, stored in a retrieval system, or transmitted, in any form or by any means, without the prior written consent of the publisher, except in the case of brief quotations, embodied in reviews and articles.

Thank you for your curiosity in reading this page. We hope you have a well-designed day that is not too hard.

ISBN: 978-0-9838731-9-8 (paperback)

ISBN: 979-8-9909793-0-7 (ebook)

Published by Berkun Media

scottberkun.com | designishard.com

TABLE OF CONTENTS

The Part Before the Other Parts **1**

Part 1: The Ego Trap **5**

Part 2: Know Your Powers **31**

Part 3: Why Design Is Hard **57**

Part 4: Make Design Easier **79**

Part 5: Design in the Real World **97**

Recommended Reading **103**

Acknowledgments **107**

About the Authors **109**

Image, Photo, and Design Credits **111**

The Part Before the Other Parts

DESIGNERS ARE LEGENDARY problem finders. This makes us good at our work, but it also makes us downers at parties: we can find something wrong with almost anything. If the host of a gathering offered us a half-full glass of wine, we very well might miss the point and think about how to redesign the glass, the tray the glass was on, and the room we were standing in. Today, many feel the design profession is in crisis, as jobs are hard to find and most people don't understand what we do. Yet others say these problems have been with us since design became a profession, so why worry now?

While we don't advise worry as a tactic (hint: it rarely helps), we do think our profession has flaws in how we think about what we do. This book offers a better way, and it starts by asking:

- Why is design so hard?
- Why are designers often frustrated at work?
- How is it designers have so little power in organizations?
- How can we be more fulfilled and successful?
- Why are our design heroes rarely helpful in showing us the way?

We are Scott Berkun and Bryan Zug, two friends, who for years have been obsessed with learning how to make better products, systems, and societies. We started meeting regularly for beers at Chuck's Hop Shop in Greenwood, Seattle, to discuss what we've learned, and this book project was born. We're here to share the highlights of hundreds of hours of research, interviews, and discussions.

In our 30 years of experience, we've worked at small startups, nonprofits, and Fortune 100 companies like Amazon, Microsoft, and Zillow. We've experienced the thrills of high-performing teams doing good for the world, and the lows of dark triad (Machiavellianism, narcissism, and psychopathy) corporations racing to the bottom of late-stage capitalism. We know that design work is intertwined with power, a topic designers often fear, and untangling this mess is a central theme of this book. As Eric Hoffer wrote in *The True Believer: Thoughts on the Nature of Mass Movements*, "Where power is not joined with faith in the future, it wards off the new and preserves the status quo."[1]

As any good designer does, we will treat your time, as our customer, as precious. This book is challenging, opinionated, and short. It is also loving, albeit often the tough kind. Our background is in UX design, project leadership, and systems, but much of the advice we offer is useful for any designer, or expert, as the patterns that make our jobs hard are similar.

---

1   Eric Hoffer, *The True Believer: Thoughts on the Nature of Mass Movements* (New York: Harper Perennial Modern Classics, 2010).

Our wish is that our advice helps you have a great career, and more importantly, a fulfilling life. Since this project began in conversation, we hope it helps you have better ones with friends and coworkers. We wish this even if you disagree with us, think we had too many beers, or were completely out of our minds for writing this book. When you finish reading, please visit designishard.com to grab our discussion guide and share your perspective so we can learn from you. Thanks for being here.

• PART 1 •

The Ego Trap

A GOOD PLACE to start a book about why design is hard is with the story of Paula Scher and her $1.5 million napkin sketch. Back in 1998, two giants in the banking industry, Citicorp and Travelers Group, were preparing for a merger. With a combined valuation of $140 billion, they were set to become the largest bank in the world.[1] Scher's firm, Pentagram, was hired to develop a new logo that would unify the brands for consumers. In one legendary meeting, while passively listening to banking executives complain about their concerns, she sketched a simple drawing on a napkin and said, "This is your logo." And she was right. Scher became a design legend.

---

1   Mitchell Martin, "Citicorp and Travelers Group Plan to Merge in Record $70 Billion Deal," *New York Times*, April 7, 1998, https://www.nytimes.com/1998/04/07/news/citicorp-and-travelers-plan-to-merge-in-record-70-billion-deal-a-new-no.html.

That rough, hand-drawn concept would lead to one of the most successful design transformations in history. Pentagram was paid $10 million for the project, and $1.5 million for the logo alone. This story is well known, as most legends are, because it's dramatic, memorable, and inspiring. However, the more important facts for designers are rarely told.

It's only the full story that expresses Scher's lesson on why design is hard. She knew the idea in the sketch, as good as it was, was just a fraction of her job. She explained that "the design…is never really the hard part…. It is persuading…people to use it."[2] Scher faced months of presentations, meetings, and iterations for dozens of different use cases to deliver that seemingly simple idea to the world. Scher knew that no matter how brilliant her idea was, the work of shepherding the idea through the organization remained. She offered, "The job was to try to get either an individual, a group of people, or

---

2   *Abstract: The Art of Design*, season 1, episode 6, "Paula Scher: Graphic Design," created by Scott Dadich, aired February 10, 2017, Netflix.

a whole corporation to be able to see…there were a million meetings trying to get buy-in." If no one did that work, the sketch would never have earned its place in design history because it would have been rejected like countless others and thrown away.

When designers look at these two images, a sketch of an idea and a finished logo actually used by a Fortune 50 company, it's easy to overlook the years of effort between them. It's understandable to look at that sketch and think, "I could make that, so what's the big deal?" As designers, we're drawn (pun intended) to surface aesthetics. However, success is rarely defined by surfaces. More often, the measure of success is the number of people convinced, the complex details resolved, or the organizational constraints overcome. History proves that the brilliance of an idea does not translate into adoption unless someone convinces people to do so.

Remember that it took a century for most of us to agree with Copernicus that the Earth orbits the sun, and not the other way around. Even today, there are still holdouts like flat-Earthers who refute these well-proven ideas. Often, the better an idea is, the harder it is to get people to accept it, since it requires them to change.[3] One of the most dangerous myths creative people have is that good ideas speak for themselves, but history proves they rarely do. Ideas challenge people's beliefs, their identities, and their habits, and they will instinctively fight hard to protect them.

In a 2023 Twitter poll (posted days before Elon Musk inexplicably dumped 17 years of brand equity by renaming it X), over 200 designers responded to our poll about what's the hardest part of making good design happen.[4] In first place, with 57.4% of votes, was working with other people. In second place, with 38.4%, was understanding problems, which emphasizes the importance of good user and customer research. In last place, with an astoundingly low 4.2% of votes, was crafting the ideas. We don't take this to mean that crafting ideas is easy, because we know sometimes it's hard. Instead, we take this result to signify that crafting ideas can be a solitary process, but bringing ideas to the world requires relationships and organizations where designers are not in control (we'll discuss this in depth in Part 3).

---

3   Scott Berkun, *The Myths of Innovation* (Sebastopol, CA: O'Reilly Media, 2010).
4   Scott Berkun (@berkun), Twitter, July 17, 2023, https://x.com/berkun/status/1681039529745199104.

Yet design culture and education focus on crafting ideas. This creates a profound, and frustrating, gap between where our attention is trained to go, and how important decisions actually get made. Even if you aren't frustrated by your career, or the state of the working world at large, we're sure you know many designers who are. These frustrations are more common for design school graduates than for people who learn on their own, but they are common nonetheless. The common reasons designers feel this way include:

- Having your ideas ignored and misunderstood
- Being involved late in important decisions
- Feeling undervalued in your organization
- Needing to explain or justify your role
- Feeling frustrated that powerful people are ignorant about design
- Feeling defeated that nothing ever changes

The first lesson of this book is that these problems are not your fault. You entered this career expecting something better, but you've been disappointed. You may have already experienced shock, burnout, and despair. Perhaps you feel limited, held back, and that few coworkers understand you. We call the cause of these feelings the *ego trap*: the belief that because you are a designer, you should be the creative hero in the story of your organization. If you feel stuck, or you know other designers who feel this way, this trap explains why. This book will teach you how to escape the trap and learn to thrive.

Many designers are in the trap now, or they've been in it before. The problem is past generations of designers failed to see the trap and failed to teach us all how to get out of it. We're sorry this happened. We wish we didn't have to write this book, but we do. We're convinced the trap is rampant and self-reinforcing, and it will continue to hurt future generations.

We admit it's possible we're wrong about this. Perhaps all designers are flourishing and throwing secret dance parties celebrating their role in late-stage capitalism, and the problem is just that we haven't been invited. We don't think so, but if we're wrong, please invite us! We promise not to redesign your home (unless you ask us to).

But if we're right, then the **second lesson of this book is that you must now change something.** Expecting the world to change to suit us is not realistic. Instead, we need to have a better mental model for how our work fits into the world. We are designers who create metaphors for our customers, but we desperately need better metaphors for ourselves.

There are three approaches to consider:[5]

- A. **Seek power.** Being undervalued means you do not have the power you need. Who makes decisions you think you should be making? Who doesn't listen to your advice but really should? Designers need power to design. There's no way around it. Decisions are really about power, and you need to increase how much power you have.
- B. **Become influential.** If you don't want the responsibilities that come with power, that's OK. Instead, become an influencer. Think of your job as a consultant or an advisor, and draw from the rich heritage of skills those roles have always had. If the powerful people you work with listened to you 30% or 50% more often, and gave you more credit, would you enjoy your career more? If yes, then influence is the way.

---

5   These approaches were explored first on Twitter in 2019, and there is some interesting commentary around these ideas. Scott Berkun (@berkun), Twitter, June 24, 2019, https://x.com/berkun/status/1143221872638586880.

C. **Be self-aware.** Even without wanting more power or influence, if you can mature your beliefs about design and escape the ego trap, you'll become a healthier person. Your career will have more flow and be more fulfilling. You'll get smarter at identifying healthy places to work, or perhaps you'll realize you want to be your own boss. By becoming self-aware, you'll be less reactive to the messy reality of human nature in organizations.

This list may scare you. Much of design culture rewards pretending there's a safer way. We've searched, and we don't think there's an alternative. Scher's story about how ideas don't advocate for themselves illuminates the necessity and purpose of this book.

If you don't like these options and prefer to wait for the world to conform to you, we wish you luck, but this book won't help you. As Anaïs Nin wrote, "It was not the truth they wanted, but an illusion they could bear to live with." We do not have to accept a *bearable* life. We are designers, gifted with creative powers few people have. It's time to use these skills to our advantage and make design easier.

Designer or Decider?

When you play a game like *Minecraft*, *Stardew Valley*, or even *Solitaire*, you are in complete control. You don't have to negotiate with anyone else. Except for the rules of the game, you have complete power and freedom, which is why video games are fun. When things go well, you get all the credit, and if things go poorly, at least all of the mistakes were yours too. Essentially, you get to design and decide almost everything. Most design education puts students in this role, where the student works alone and makes all of the design decisions.

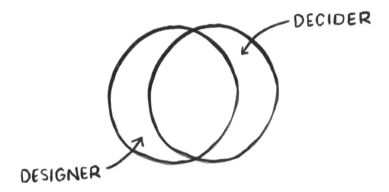

How does this compare to the average professional designer? If your job title has the word *design* in it, what do you actually decide? What can you choose to do without having to convince someone else first? Probably not many things, if any at all. While you're free to conceive ideas, you're often almost the *opposite* of free in how those ideas are manifested into the world. You may frequently feel like a victim, perhaps of politics, bureaucracy, ignorance, or short-term thinking. While you can conceive anything you want in Photoshop, Figma, or on a sketchpad, it is the powerful who decide whether it gets created (unless perhaps you're an entrepreneur[6]).

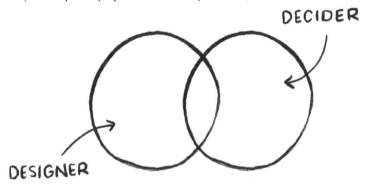

---

6   Designer Fund provides one model for how designers can be organizational leaders or company founders. See https://www.designerfund.com/.

Somehow we've forgotten that we often define *designing* as separate from the act of *deciding*. What we call designing is mostly *advising*. We make things as suggestions for other powerful people, clients, executives, and project teams, and they make the decisions. We often get to decide only small and inconsequential details. We work hard to offer good advice but that advice is often ignored. This means the true designers of a product are not the people with the job title *designer*, but the leaders who decide goals, staffing, and budgets. As designer Maxim Leyzerovich cynically described, "Being a designer is mostly just making people feel good about decisions that have already been made."[7] But what would it mean to be a designer who is also the *decider?* What would have to change?

This situation is not what you expected when you started your career. You likely imagined teams of people, minions perhaps, who'd joyfully work to bring your ideas to life. You did not expect to be treated like a minor character, spending your time making bad products pretty (which can be impossible to do), or watching people with little design knowledge define entire products, strategies, and platforms while ignoring your talents. This shock is compounded by the fact that no one else in the organization seems to notice this problem, and when you try to point it out, they're confused or defensive. How could the world be this way? It's a common question for talented designers to ask, but it's the wrong question. A better question is: why would it be any different than it is?

It helps to take a wider view: we just don't think our species is all that wise. Humans aren't that good at civilization, much less running organizations. We're very self-involved, and we don't learn from history,

---

7 Maxim Leyzerovich (@round), Twitter, February 10, 2022, https://x.com/round/status/1491809762614431747.

repeating the same mistakes, like starting foolish wars or squandering resources. Just look at how we're (not) handling the climate crisis we've created on our planet, something all 8 billion of us should agree to prioritize. Taking a step back, it's easier to see that people in large groups often do foolish things. Many companies go out of business or aren't run very well. Maybe this is just the best we've been able to do so far?

Part of the ego trap for designers is assuming any problem we face is targeted at us. It makes more sense that we're just along for the ride, and the frustrations we feel are collateral damage in a bigger story. In writing this book, we believe there's much to be proud of and hopeful for: it's better to be alive now than 100 or 500 years ago. But at the same time, it helps to realize some things are broken and unfair on a large scale. Plenty of other professions, including school teachers, healthcare workers, and other underpaid but critical roles, have pressing complaints too.

It's also true that the ego trap isn't just for designers. Most people suffer from it to a degree, as each of us is the central character in our own mind. Cognitive biases like the **egocentric bias** and the fundamental attribution error prove that we're all prone to feeling special and privileged, especially in cultures that favor individualism, like the U.S.[8] However, the role designers play in most workplaces, combined with our tendency to see our work not only as a job but also as our identity, makes the trap more dangerous for us.

The root of this problem for designers is a failure of expectations. The ego trap is perpetuated in design school, culture, and media by always making the designer the hero. Designers are presented as the star

---

8 "Egocentric bias," Wikipedia, https://en.wikipedia.org/wiki/Egocentric_bias.

of every course, book, lecture, documentary film, or article about a famous project. At events, invited speakers are usually designers or people who are design-friendly, admiring and understanding the profession. And designers are often portrayed as solitary visionary artists, like Frank Lloyd Wright, rather than collaborative leaders on interdependent projects, which is far more common. Early on, the expectation is set that we will, by default, be the center of attention.

When designers make diagrams about how design decisions or UX processes should fit into an organization, design work is usually at the center, much to the dismay of project leaders, engineers, executives, and sometimes even customers. In most organizational cultures, the natural stars are the jobs that have been there longest and have played a central role, like business or engineering. Design often begins on the periphery of organizations, which is not a surprise to anyone except us designers.

The raw numbers tell us that designers are a professional minority. There are roughly 400,000 professional UX, web, and graphic designers in the United States.[9] With a working-age population of 130 million, we are a mere .03%. These numbers make it obvious that designers should be trained early on to be comfortable and confident working with people who know little about design, since that's almost everyone. Yet design culture and the ego trap have taught us the opposite: we're easily frustrated when people don't instantly see our value or speak our language.

---

9   The most reliable data we found was from the U.S. Department of Labor, which lists 110k UX Designers and 212k Graphic Designers in 2023. We've rounded up because we believe these numbers underrepresent the profession. See https://www.bls.gov/oes/current/oes151255.htm and https://www.bls.gov/oes/current/oes271024.htm.

Many of us claim to be experts at psychology and empathy, but we fail to realize how often our behavior works against us with coworkers. It's not as if our well-earned stereotypes of black clothing, intimidating obsessions with style, and furtive glances in meetings have helped us gain influence. To be clear, we feel strongly that no one should be judged because of their appearance, background, or other reasonable preferences while at work. However, that doesn't mean we aren't judged or that our choices don't affect others.

Some of our self-involvement is an understandable defense against the abuses we've experienced in our families and workplaces. When you've done the right thing in previous jobs and had your trust broken, it's hard to forget. It's also clear that the personality traits that lead to earning the label "creative" have unfortunate consequences, since human nature is fearful of people with different ideas. But defenses are also a kind of mask: they help us feel safer, but they come with a cost. As Gordon MacKenzie explained in *Orbiting the Giant Hairball: A Corporate Fool's Guide to Surviving with Grace*:

> Masks have real social value in that they allow you privacy and space in an often brutal world. But there is a price you pay for wearing a mask. Masks cause little deaths—little soul deaths. When you wear a mask, nobody (not even you) gets to find out who you really are. When you wear a mask, nobody (not even you) gets to find out what you really need. And when you wear a mask, nobody (not even you) gets to find out what you really have to offer.[10]

---

10 Gordon MacKenzie, *Orbiting the Giant Hairball: A Corporate Fool's Guide to Surviving with Grace* (New York: Viking, 1998).

Believing we are creative heroes encourages us to wear masks. Think about how many superheroes wear disguises. Masks seem cool in movies, but even in fictional worlds these heroes are usually miserable and conflicted, and the mask only separates them from who they're trying to save and protect. The masks, so poorly designed that the audience is never fooled, are mostly effective at keeping something hidden from the wearer, often a profound and terrifying truth (typically unresolved childhood trauma[11]).

For designers, the ego trap often fuels a self-serving attitude (much like a mask). We expect that somehow, magically, the majority of leaders will have knowledge only designers have, despite how they've successfully done their jobs for years before our arrival. When we encounter people who don't appreciate our skills, we passively blame them for their ignorance, often in backchannels where they can't even ask questions. We might be afraid to behave differently, but we have to admit we're contributing to patterns that limit us.

To get to the point, we've cultivated unrealistic, unspoken, and incorrect expectations about most workplaces, including:

- Design skill is already valued and leaders know how to leverage it
- Everyone knows the basics of good design principles
- There is a healthy process for making project-level decisions
- The organization is free from chronic cultural and political tensions
- People will readily give up the favorite parts of their jobs to us
- People will readily give up their sense of heroship and its rewards

---

11 Robin Rosenberg, "The Psychology Behind Superhero Origin Stories," *Smithsonian Magazine*, February 2013, https://www.smithsonianmag.com/arts-culture/the-psychology-behind-superhero-origin-stories-4015776/.

- Leaders will quickly trust us with strategic decisions
- Old, self-reinforcing, and biased systems of incentives will change upon our arrival

There is no workplace that meets this standard. Yet design culture pretends this list is universal because we aren't trained for how to do our work when they're absent. As Erika Hall, co-founder of Mule Design, wrote, "The work is so much about building relationships, analyzing power dynamics, and having the right fights. I'm not sure people are being taught this in the design schools and bootcamps."[12]

For example, Lisa deBettencourt, founder of the design firm Forge Harmonic and co-founder of IxDA, faced this situation early in her career. As a designer at a consumer electronics company, she was shocked by how much resistance there was to using her skills. The VP of Engineering consistently challenged her in meetings over the most basic design principles. deBettencourt later learned, after struggling to understand what she was doing wrong, that it wasn't her fault: product design was his favorite part of the job, and he didn't want to give it away.

It's no surprise that powerful people can be self-involved, or that workplaces are often dysfunctional, with childish politics, stupid groupthink, sexism, racism, or passive-aggressive factionalism. We watch shows like *Succession*, *Game of Thrones*, or *Yellowstone* and are well-versed in how complicated workplace drama can be. But the ego trap perpetuates the illusion that when we arrive, our design mojo will surround us in a magic bubble of butterflies and unicorns, transforming our coworkers into the best versions of themselves, specifically in ways that are the most convenient for us. This is a self-destructive

---

12 Erika Hall, LinkedIn, https://www.linkedin.com/posts/erikahall_the-entire-field-of-ux-needs-to-be-rethought-activity-7056639902469259265-zemw/.

fantasy that is rarely challenged. deBettencourt learned a critical lesson: to first investigate what it is that powerful people really care about and why. She learned it the hard way, and we don't want you to have to do that too.

Among its many sweet poisons, the trap assumes that all organizations have healthy, collaborative cultures by default. It is never explained that most organizations, like families, are dysfunctional in at least one major way, and that design—as a planning discipline that requires another role like engineering to actually build its ideas—is directly impacted by these dysfunctions. If we don't prepare for them, we're prone to learned helplessness: we'll be unable to do our work until someone else creates the conditions listed above for us, conditions we won't be comfortable identifying, mitigating, or solving on our own. To make design easier, we have to see these as solvable problems. Other professions have solved them, and we can too.

Talent Is a Distraction

We know that many designers feel underutilized at work. We also know we love to complain and see the flaws all around us. We rant about the confusing UX in the apps and websites we use, and the failings in our public transportation, school systems, and governments. We critique the bad typography on city streets and on restaurant menus. Yet we rarely ask, why is this the state of things?

This leads to two questions:

- Why aren't designers better utilized in the world?
- What are we doing to change this?

The incorrect answer to these questions is that there's a lack of design talent (which we define as a combination of knowledge and skill). This is not the problem, since so many of us know we're not used effectively. Talent is only useful if the person who possesses it has either power or influence, and we typically have a shortage of both.

For example, think of all the meetings where your best ideas were ignored. If you were two or three times as talented, would it have made any difference? Probably not, because the people you needed to convince lacked the understanding of good design necessary to *value* your talent. However, if you had twice as much power, you might have been the decision-maker and wouldn't need to convince anyone of anything. Or if you had three times as much influence, they would have trusted you, even if they didn't fully understand your ideas.

On the first day of design school, and in the first page of every design book, there should be a clarifying lesson about human nature. Decision-making in the real world is a social process, not a solitary one. This means design is usually a social process too, even if it's just between a visionary solo designer and their fancy client. So we suggest:

> Your success as a designer depends **as equally on your relationships as it does on your design talent.**
> The powerful people you need as allies, or your clients, will likely know little about design, and you will have to teach and persuade them. Your amazing ideas and concepts can't help the world if they are never built by your organization. Your ability to explain your design ideas and convince people to use them is equally as valuable as your creativity.

INFLUENCE BEATS SKILLS EVERY TIME

This means that influence and power are the multipliers you need. This explains why someone with low design talent but great power (like a CEO) can do terrible things, whereas someone with great design talent but low power likely has no impact at all.

For many design purists, this is a betrayal. For them, design is strictly the crafting of ideas, which means skills like persuasion should not be considered part of a creative profession. A purist might say that if they wanted a job based on relationships, they'd become project leaders, product managers, or entrepreneurs (which we explore in Part 3). If their design heroes succeeded based on solitary use of their creative talents, why can't they? And this is where the dangers of hero mythology strike again.

Dieter Rams is one of the legends of modern design, famous for his influential work at Braun that established the styles and principles used in Apple's most successful products, like the iPod and iPhone.[13] As with many design heroes, Rams is often pictured standing alone in front of the dozens of famous products he designed.

The danger is that hero shots of designers like Rams encourage us to presume that he had teams of minions bringing his ideas to life, and all designers should aspire to this. Remember how the simple image of Scher's famous $1.5 million logo obscures the two years of effort required to bring it to the world. This image of Rams falsely conveys that the hard part was crafting the ideas (i.e., 4.2%) and that

---

13 Jacopo Prisco, "Dieter Rams: The Legendary Designer Who Influenced Apple," CNN, November 20, 2018, https://www.cnn.com/style/article/dieter-rams-film-exhibition-style-intl/index.html.

there were no collaborators, financiers, politics, or friction along the the way.

Fortunately for us, Rams revealed how relationships with non-designers (a term we avoid, as it is most humans) were critical to his success. While developing consumer electronics at Braun, Rams realized there was dysfunction between the engineering team and the design team. Rather than avoid the problem, or blame someone else, he decided it was part of his role as a designer to do something about it. He explained, "I noticed the engineers liked brandy, so I'd buy a bottle of good Cognac to share. To be a good designer, you have to be half-psychologist."[14]

Rams is literally telling us that half of his job was building relationships, which is how he achieved the work he became famous for. This explodes the ego trap. For that reason, you won't hear stories like this in many design books, courses, or documentaries. You won't find any photos of him with the many engineers, project managers, business analysts, or marketers he collaborated with or took orders from. The legendary bottle of brandy should be stored behind glass at the MOMA next to Braun's works, and pictured in every design book, because it represents design as a *relational activity*, not a solitary one. It should be obvious that building relationships involves much more than buying gifts, but it's a good start.

As design historian Alice Rawsthorn wrote in *Hello World: Where Design Meets Life*:

> Any number of factors can impede the progress of a design project: the unwelcome intervention of colleagues,

---

[14] Alice Rawsthorn, "Reviving Dieter Rams's Pragmatism," *New York Times*, November 12, 2006, https://www.nytimes.com/2006/11/12/style/12iht-design13.html.

clients, or suppliers; budgetary restrictions; human error; and sudden changes that the designer could not have anticipated, and cannot control. The completeness with which any individual designer's vision will eventually be realized is usually determined by a combination of their strength of character, persuasive powers, and luck.[15]

We'd emphasize relationships more than Rawsthorn does here, but the wider spirit of her point is echoed by most experienced designers we know. Yet because these insights conflict with the ego trap, they're not considered part of being a designer. We know it's unlikely you've heard these stories about Scher and Rams. Instead, our popular stories fit the trap, motivating young designers to chase a fantasy that even our heroes deny. That's how dangerous hero myths are.

Use Productive Idealism

Early in this book we shared that designers have the power of *skepticism*: we are gifted with the ability to find flaws in things so they can be fixed. A more positive way to look at the same ability is by labeling it as *idealism*. We are positive that everything can be improved—a powerful belief fueling the way we look at the world. We sometimes forget that no sane person hires a designer, or becomes one, with the goal of keeping everything exactly the same (although sadly that sometimes happens anyway). The risk, however, is that idealism can lead us into self-limiting arguments that work against progress and frustrate everyone.

The warning sign that idealism has gone too far is the use of the word *should*. This is a dangerous word because it presumes a convenient,

---

15 Alice Rawsthorn, *Hello World: Where Design Meets Life* (London: Hamish Hamilton, 2013).

super-powerful authority is waiting around to agree with us, and all we have to do is reference them to get our way. Classic examples include statements like, "design should have a seat at the table," or "UX should be involved earlier in the process." The statements are popular because they feel good to say (and because they might be true in our belief system). The problem is that should statements are fallacies and have little effect on other people.

The reason they're fallacies is that every powerful person in your organization has their own *shoulds*. Their shoulds are behind all the decisions they make, including the ones that frustrate you. Since these other roles have more power than you, their shoulds will always "out-should" yours. That's how power works. As David Graeber wrote in *The Utopia of Rules: On Technology, Stupidity, and the Secret Joys of Bureaucracy,* "Power is all about what you don't have to worry about, don't have to know about, and don't have to do."[16]

If you don't have enough power to decide, then you must use your influence to change their minds, and should arguments just aren't persuasive. Changing people's minds, as challenging as it is, must therefore be a skill any person serious about their ideas needs to learn.

The question then is: why do we hear so many should arguments? One reason is that if a person is influential, a should argument earns attention from others out of respect and trust, even if it goes nowhere. Another is that should arguments are emotional, and in the bland landscape of prioritized action items and endless threads in communication apps, people sharing actual feelings get attention regardless of the merit of the argument.

---

16 David Graeber, *The Utopia of Rules: On Technology, Stupidity, and the Secret Joys of Bureaucracy* (Brooklyn: Melville House, 2016).

But the more common answer is that should statements rally support from other frustrated people who share your preferences. It's preaching to the choir. For people who are afraid to speak up, it's empowering to hear someone else say what they are thinking. Design books, conferences, and events often have many convenient arguments about how the world should be (i.e., the ego trap). They get applause, but anyone who has good counterarguments isn't likely to be reading these books or attending these events.

The better approach is to admit that should statements rarely change minds because they don't explain why *your should* is better than *their should*. To be persuasive, **you have to express why your should is better than their current position.**

For example, these are unproductive, idealistic statements:

- I should be allowed to <*insert thing you are not allowed to do*>
- They should never do <*thing they are rewarded for doing*>
- Why should I have to <*insert thing you don't like*>

These can be easily converted into productive, persuasive arguments, but you have to do the homework. For example:

> I should be in this meeting because I have proven
> to you leaders, Aya and Juma, that I help facilitate good
> decisions. I know you need to keep the meeting small,
> but as a designer, I can quickly sketch ideas to help
> us understand their value before we make decisions,
> which will make the meeting more effective for everyone.

This gives the powerful person a comparison to consider. Important, busy people love comparisons (is A better than B?). You do need to prepare and have self-awareness (i.e., are you perceived as facilitating good decisions?). The payoff for doing the homework is that by arguing in their self-interest, their ego and goals become an asset to you—at least part of them wants to argue your side. And even if they don't agree, you set yourself up to politely ask, "Thanks for considering this. What can I do to prove enough value in the future to be included?" If they say anything useful, you can come back in a week or month with that proof and ask again. We admit the details here may vary widely, but our point is that should statements alone have none of these advantages.

To further explore the danger of shoulds, here's a fictional but common situation (that, to no surprise, magically supports all of our arguments in this book so far):

> Kailani is a third-year product designer. She's the only designer in a meeting for a web redesign. Marketing wants something "that pops." The PM demands 20% Daily Active User growth. The VP interrupts with a sketch for an idea based on a dream her spouse had last night. The engineers start generating dev estimates. Then everyone leaves. The PM asks Kailani to mock something up. Kailani is upset, with little faith in this project.

This is an unfair situation in many ways, but it's not uncommon. To our point, little in Kailani's training prepared her for this. She was taught to base her confidence on her UX design skills, not situational awareness or her relational abilities, like how to facilitate difficult conversations. And she is in the more difficult position of being the most junior person in the room and likely the only woman and person of color. She doesn't know who her real allies are or how to enlist their help. If she had a good manager they would have prepared her for this meeting. And any good leader in the room should have noticed the situation and supported her. But we know the problem with should statements: what should happen often doesn't.

At happy hour with her design coworkers, Kailani shares her story. Immediately, they all commiserate about this common experience.

"They should know better by now," one says. "We shouldn't have to deal with this," says another. "This is the way it has always been," someone might say, which is not entirely wrong. There's bonding and validation in these conversations, which is valuable for her mental health, but without real-world solutions, complaining is ultimately a disempowering path. The harder truth lurking in this story is that the ego trap, which implies others are rejecting her personally and explicitly, makes her a design victim, since this framing is rooted in unrealistic expectations.

Designers often say "they just don't get it" about organizational leaders, as if "getting it" is a realistic expectation for people who are currently rewarded *not* to get it. It's like being angry at a rock for being so quiet and unmoving. Only us designers remain mystified. We keep hoping someone will magically appear to save us from the responsibility of persuading people to see the value in our skills and contributions. But who exactly could do that? If our boss understood our value, of course they should do that for us, but what if we don't work for that kind of person?

Productive idealism suggests that instead of saying "they should" or "they don't get it," *we* are responsible for asking better questions of ourselves or the people we're trying to persuade:

- Who exactly should know better?
- How are they benefitting from not knowing?
- What incentives would motivate people to know?
- Who controls those incentives? Why would they change them?
- Who gets it the most? How can we enlist their help?
- How can we make our expectations more realistic?
- What is the first, smallest, easiest thing to get them to see?

We're inspired by Scher and Rams for their uncommon honesty in talking about the relational nature of their success. True design heroes anticipate problems. They study the organizational landscapes they're in. They train for what *usually* happens, rather than for what they *hope* happens. If nothing else, they help us see our choices. We don't have to do extra work or tasks we won't be rewarded for, but we do have to find the maturity to see the real reasons why we're frustrated, and either accept them or proactively solve them. This is the focus of the rest of this book.

• **PART 2** •

Know Your Powers

THERE IS A dark truth lurking beneath the waters of Part 1. For your safety, we must reveal it now, in the seemingly calm waters of Part 2, so it doesn't surprise you later. Please refrain for the moment from juggling knives, disarming bombs, or other dangerous activities where surprise could result in instant death. We don't want you to die, especially not while reading our book, since others might see it in your hands postmortem and that would be bad for sales.

Here's the tough truth: no one escapes the ego trap! Were you expecting something worse? Hopefully you're relieved. The reason no one escapes the trap entirely is because we need our egos. The human mind depends on them. It's useful to place ourselves at the center of the story, at least some (or let's be honest, *most*) of the time. Creative work depends on ego, but it has to be used in ways that help you achieve your goals, rather than work against them.

Sometimes we need to complain about our situation to get through the day or to restore our sanity. Appealing to entropy, the awareness that life is chaotic or unfair, is cathartic. In some moments, saying, "they should know better," or "the system is broken," or "we were set up to fail" is 100% true. Even when it's only half-true, saying it is can be validating and relieving, which are important emotions to feel. There are good reasons why much of the greatest art in history is about heartbreak (especially common in popular music) or despair. An example from our "entropy / everything is falling apart" playlist is Thom Yorke's "Black Swan." Its powerful lyrics and driving, syncopated rhythm help us feel understood:

> What will grow crooked, you can't make straight
> It's the price that you've gotta pay…
> People get crushed like biscuit crumbs
> And laid down in the bitumen
> You have tried your best to please everyone
> But it just isn't happening
> No, it just isn't happening
> And that is fucked up, fucked up[1]

When we're having a rough time, songs like this help us sort through our feelings so we can move past them. Here's another song on our entropy playlist, one with a more positive view of existence, "Fight for Love," by Sault:

> How do you fight for love
> When the world has broken you?
> I know that times are rough
> When you're down, keep looking up[2]

---

1 Thom Yorke, "Black Swan," track 4 on *The Eraser*, 2006, XL Recordings.
2 Sault, "Fight for Love," track 7 on *11*, 2022, Forever Living Originals.

Even if you don't like our music choices, you must embrace art that helps you understand your emotions. Mastery of your creative powers comes from self-awareness, whereas the ego trap is mostly denial. Ego hides other feelings necessary to create good things for the world. If you linger in self-righteousness, anger, and disappointment, there are many great ideas you'll never find.

One good definition of what designers do is repairing what's broken. Another is bringing light to darkness. A third is finding creative escapes from perceived dead ends. Regardless of how you define it, design is not a good career for people who only want to say, "this can't be done," or "this won't work," or "this organization is hopeless." Anyone can say those things—they require no effort or creative talent.

Paula Scher didn't say, "To hell with these bankers and their corporate bureaucracy" and quit, as fun as that might've been (as an aside, we don't think helping megacorporations like global banks is good for the world, but more on that in Part 3). Rams didn't decide, "This brandy is too good for these Neanderthal engineers, and I don't like schmoozing," even if those things were true. As Anaïs Nin said, "We don't see things as they are, we see them as we are." When we're ready to give up, it's only by shifting our perspective that we'll see the way forward.

The more direct way to make this point is: do not waste time complaining about **gravity problems**. We don't spend our time complaining about gravity, and how heavy many things are to lift, because it is an immutable law of physics. Complaining about gravity is about as useful as complaining about why time only goes forward or why dirt doesn't taste like chocolate. In books by the Stoic philosophers like Seneca and Marcus Aurelius, they repeatedly hit readers over the head with the advice that life is better when you don't dwell on what you can't control—a lesson designers desperately need to learn.

The term gravity problem comes from Bill Burnett and Dave Evans, in their book *Designing Your Life: How to Build a Well-Lived, Joyful Life*. They don't pretend that big problems don't exist or that you should avoid them. Instead, they offer an excellent definition of what wise designers do:

> The key is not to get stuck on something that you have effectively no chance of succeeding at. We are all for aggressive and world-changing goals. Please do fight City Hall. Oppose injustice. Work for women's rights. Pursue food justice. End homelessness. Combat global warming. But do it smart. If you become open-minded enough to accept reality, you'll be freed to reframe an actionable problem and design a way to participate in the world on things that matter to you and might even work.[3]

To make this clearer, we think wise designers:

- Have open minds
- Are rooted in the real world
- Reframe (or redefine) important problems
- Make problems actionable

Designers are natural reframers. That's what we're doing when we're experimenting with different page layouts for a poster, or asking, "What's the real problem we're trying to solve?" during a chaotic brainstorming meeting. We just need to apply these reframing skills to the harder problems we face in our jobs.

Let's try this out with a common situation designers face:

---

3   Bill Burnett and Dave Evans, *Designing Your Life: How to Build a Well-Lived, Joyful Life* (New York: Alfred A. Knopf, 2016).

> Jesse finishes her presentation for the website redesign to the leadership team. They applaud and she smiles. One executive, Hana, immediately raises her hand. "This was great, loved it. But why is the primary color green? It should be cornflower blue, because… *<insert strange anecdote expressing what is at best a mild personal preference that in any sane world is comically insufficient as a legitimate argument for a decision when compared to Jesse's design expertise conveyed in her thorough presentation>*."

If you've been in this situation, as most designers have, it's embarrassing and disappointing. Jesse is seemingly forced to justify years of training and experience, which is insulting. This fuels the *shoulds*, like "they should know better" or "they should just trust me," but we know the problems inherent in shoulds. Instead, let's apply our latest definition of design (i.e., use an open mind and be rooted in the real world) to reframe the problem.

With an open mind, we can list these real-world possibilities for what's happening:

- Many powerful people won't know anything about design
- We might be the first (good) designer they've talked to
- We will often get asked basic questions based on low trust
- Meetings can be theater, and people speak up for strange reasons
- Some people in the room are probably on Jesse's side

This list changes the problem we need to solve. This makes Jesse's situation less of an insulting surprise and more of an expected challenge. We just have to admit that this scenario is common and an unfortunate part of the job, instead of betting on the false hope it won't happen. Then we can study it, craft ideas around it, and solve it. Perhaps most surprising is that by framing the situation this way, the powerful person, Hana, actually has a major disadvantage: she's wandered onto the designer's turf, where we have far superior knowledge.

In Susan Weinschenk's book, *100 Things Every Designer Needs to Know About People*,[4] she explains the famous Dunning-Kruger effect. Put simply, it's a cognitive bias that people with less knowledge on a subject tend to be overconfident about how much they know (e.g., 65% of people think they are of above-average intelligence, which is impossible[5]). All experts, including pilots, doctors, and lawyers, contend with ignorant questions fueled by the Dunning-Kruger effect. Doctors treat patients every day who, after only skimming a handful of internet search results, will argue about basic medical knowledge. Just like Jesse, scientists, politicians, and even executives, frequently encounter astoundingly foolish questions from people who do not have their expertise.

If we accept this as a gravity problem, simple solutions include:

- Practice the presentation with a non-designer who asks tough questions
- Prepare a back-up slide with brand research, moodboards, and rationale for the color scheme, explaining why this approach is superior to personal preferences
- Ask a powerful ally beforehand to be ready to speak up during the meeting
- Improve public speaking and improvisational skills for general confidence

---

4   Susan Weinschenk, *100 Things Every Designer Needs to Know About People* (Indianapolis: New Riders, 2011).
5   Patrick R. Heck, Daniel J. Simons, and Christopher F. Chabris, "65% of Americans Believe They Are Above Average in Intelligence: Results of Two Nationally Representative Surveys," *Plos One*, July 3, 2018, https://journals.plos.org/plosone/article?id=10.1371/journal.pone.0200103.

This might sound like extra work, and you might be thinking, "Why should I have to do this?" The answer is you might not have to. Maybe it's OK with you to live in a cornflower-blue world. Or you feel you should be paid more if you have to fight against clueless executives. Perhaps you've found a boss who's a design champion and protects you from these situations. Those are all reasonable choices.

However, odds are you will face this situation in your career. When you do, it becomes clear that you are not the decision-maker. If you were, someone would be presenting to you for your approval. If you are presenting, then you are trying to influence a decider, and the list above will help you.

When done well, influence work becomes easier as people learn basic design knowledge from you and grant you more trust. This is what design leaders do: they pave the way in organizations so fellow designers will experience less friction and face fewer gravity problems.

Powerful Roles to Play

Reframing is one of many powers you possess that can make design easier instead of harder. In Part 1, we explained that designing and deciding are often different roles. If designing isn't deciding, what else is it? Deciders are busy, so we have to take responsibility for providing answers. Too often we expect our job title to do the work. We expect powerful people to figure out what value we provide, learn our fancy terms, and know the difference between a half-dozen design specializations. Instead, your curmudgeonly authors agree with Norman Potter's classic book, *What Is a Designer: Things, Places, Messages*, where he explained that "the words by which people describe themselves—architect, graphic designer, interior designer, etc.—become curiously more important than the work they actually do...."[6]

When you hire a good plumber or an interior decorator, they show up and demonstrate their value. They aren't insulted if you don't know their terms or methods because they assume you know nothing. They treat your time, as the decider, as precious (at least before the contract is signed). They take responsibility for helping you understand the problem and how their expertise can solve it.

Whereas in our world, we habitually shift through different names (UX designer, interaction designer, user researcher, user interface designer, etc.), hoping that one day there will be a magic term that does the explaining for us. Unless you're sure a new name or term helps you, perhaps because it gets you more attention from executives or is familiar to the engineering team, it's a waste of time. Why? Deciders don't care about your title—they care about how you can solve their biggest problems. That's it. It's not that complicated.

If you can go to a decider and authentically say, "From your list of important unsolved problems, I can solve #2, #3, and #4. Here's what I need from you to do it," you will have their full attention. You will probably get budget and resources too. Only then are you speaking their language instead of obsessing about your own.

In *Articulating Design Decisions: Communicate with Stakeholders, Keep Your Sanity, and Deliver the Best User Experience*, author Tom Greever explained that:

> The difference between a good designer and a great designer is the ability to not only solve the problem,

---

[6] Norman Potter, *What Is a Designer: Things, Places, Messages* (London: Hyphen Press, 2002).

> but also to articulate how the design solves it in a way that is compelling and fosters agreement...
> There are three things that every design needs to be successful: it solves a problem, it's easy for users, it's supported by everyone.[7]

To us, the good designer Greever describes is an ineffective one. Why? That designer will often leave his best work on the drawing board, because without agreement or support, no customer will ever see it. Chuck Harrison, one of the first black designers and executives in U.S. history, who designed hundreds of successful products for Sears, expressed a similar discovery he made midway through his career: "I learned [design] wasn't only expressing a concept, but how to successfully get it to the marketplace."[8] The good news is you already have many of the skills to do this. All that's needed is to reframe them so they can be used in a more powerful way.

---

7 Tom Greever, *Articulating Design Decisions: Communicate with Stakeholders, Keep Your Sanity, and Deliver the Best User Experience,* 2nd ed. (Sebastopol, CA: O'Reilly Media, 2021).
8 Charles Harrison, *A Life's Design: The Life and Work of Industrial Designer Charles Harrison* (Chicago: Ibis Design, 2005).

Investigator

The best designers are naturally curious. They always have questions, and then questions about the answers to those questions. We don't accept things at face value: we want to understand how things really work, and why they are the way they are. We want user research, market research, web analytics, and any source that might help us answer our questions. We know that data is a flashlight that helps us see the truth, and we want to design for the world as it really is. If we can spend hours reading about the 16th-century French history behind the beloved font Garamond, or studying the details of the design prototypes Jonathan Ives made to create the first iPhone, we have the rare capacity to discover and digest layers of complex information for practical use in solving problems.

Our investigative curiosity is magical in two ways. First, we can use it to find better answers to important questions organizational leaders have. And second, we can aim our investigative powers at things that we find annoying, like dismissive coworkers, dumb bureaucratic processes, frustrating clients, and short-sighted executives. Ted Lasso said, "Be curious, not judgmental,"[9] and to that we'd add, it's only by being curious, rather than judgmental, that we make discoveries toward reframing our frustrations. Instead of obsessing about not having a seat at the table or why our ideas get ignored, reframe the problem. Make a habit of asking questions like:

- What are the most important unsolved problems leaders have?
- Which powerful people have problems I can solve, but they don't know it yet?

---

9   *Ted Lasso*, season 1, episode 8, "The Diamond Dogs," developed by Jason Sudeikis, Bill Lawrence, Brendan Hunt, and Joe Kelly, aired September 18, 2020, Apple TV+.

- How can I sell them on a small, safe project to demonstrate my value?
- Who defends the status quo? Who wants change and how can I ally with them?

As Ellen Lupton, famed graphic designer and curator, said, "Design is an art of situations. Designers respond to a need, a problem, a circumstance, that arises in the world. The best work is produced in relation to interesting situations."[10] Be an investigator of the project you're on. Be curious about team leaders. What frustrates them? Which deciders get along and which ones don't? By playing the role of investigator on the totality of your project, not just the narrowest scope of design, the design work itself becomes easier.

The bad news is that there's a natural gravity problem around who gets to work on investigative projects, like future planning or long-term strategy. Such projects are scarce and highly sought after. Powerful people typically assign the responsibility to those they trust the most. This is rarely designers, because we usually haven't been in the organization as long, so we haven't been granted (or earned) enough trust. The way forward is to rely on curiosity and the list of questions above.

Explainer

The biblical Tower of Babel fell not because of failed design or engineering knowledge, but because the people building it lost the ability to understand each other. The same problem explains the dysfunctions of many organizations. The working world is an overwhelming

---

10 Ellen Lupton, *Thinking with Type: A Critical Guide for Designers, Writers, Editors & Students*, 2nd ed. (New York: Princeton Architectural Press, 2010).

sea of jargon and doublespeak, spoken and written words that drown productivity and morale. Even smart, talented, well-intentioned people end up confused, disorganized, and demoralized. Expensive productivity apps and communication tools are perennially promised as the solution, but it's a fantasy silver bullet because the real problem is social—a lack of communication skills and common language—not technological.

This means someone who explains things clearly, including through insightful sketches, diagrams, or metaphors, has tremendous value. Explainers help people make sense of each other. Designers are often shy about their ability to explain things, but typically we're better at this than other professionals, since our work is rooted in communication (even visual design is rooted in semiotics, the study of symbols and their meaning). If we can be curious about our coworkers' perspectives, objectives, and frustrations, we can be translators. Translators have power: they can become the key link in how project teams function. Good leaders depend on their translation skills to earn trust and help better ideas arise. Designers can also do this, and when we do, our jobs become easier.

In the book *Communicating the UX Vision: 13 Anti-Patterns That Block Good Ideas*, authors Martina Hodges-Schell and James O'Brien asked designers to interrogate their own communication breakdowns:

> Think back over your latest project and see if there are moments of disagreement or circular conversations that could be explained by this mismatch of semantics. Were there moments where you dismissed or failed to understand a concern because it was in a different dialect?[11]

---

11 Martina Hodges-Schell and James O'Brien, *Communicating the UX Vision: 13 Anti-Patterns That Block Good Ideas* (Waltham, MA: Morgan Kaufmann, 2015).

They call out the fact that in design education, little time is spent on communication skills, even though in our work as advisors, our communication matters more than our design skills (as we explained in Part 1). We have to communicate with stakeholders before we design anything, or even after, to get their approval. Hodges-Schell and O'Brien add:

> As designers, we have the perfect toolkit to collect the dialects of our organization, translate them into a simple visual form everyone can understand, and capture them in a common vocabulary for the life of the project.

It's worth emphasizing that visual explanations play to our strengths, and science supports our unique powers. Images are processed in older parts of our brains than text (written language is a very recent invention), which is why humans process visual information faster. As education expert Lynell Burmark explained:

> Unless our words, concepts, ideas are hooked onto an image, they will go in one ear, sail through the brain, and go out the other ear. Words are processed by our short-term memory... Images, on the other hand, go directly into long-term memory where they are indelibly etched.[12]

Below is what should be a familiar example, presented in two different ways. There's a palpable difference in the experience of consuming a concept through visuals than through text. Explaining things well

---

[12] "Why Data Visualizations Help You Discover Meaningful Insights from Data, Faster," LexisNexis, July 27, 2022, https://www.lexisnexis.com/blogs/sg/b/data-as-a-service/posts/why-data-visualizations-help-you-discover-meaningful-insights-from-data-faster.

visually is satisfying in a way that's hard to describe but everyone recognizes.

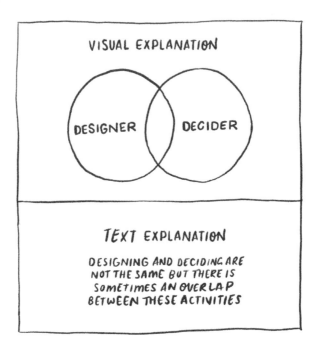

If you're ever the only designer in a meeting, see it as an advantage. If you listen for points of confusion and volunteer ways to reduce it, people will love you. During a difficult meeting, where people are arguing and not listening to each other, go to the whiteboard, draw a quick sketch, or even a 2x2 table of the possible pros and cons, and say, "Do you mean this or this?" Often, magic happens. From your prompts, people start using parts of their brains that were sleeping, the parts better suited for metaphor, creative thinking, and collaboration. Even if they say what you drew was wrong and correct it, the meeting has been transformed, thanks to your initiative. They may at first see your contribution as simply making things prettier (which can be an achievement too), but pay attention to who notices the deeper problem you solved: they are your future allies.

Negotiator

When you explore different ideas for how to solve a problem, what is the conversation like in your mind? From Bryan Lawson's excellent book, *How Designers Think: The Design Process Demystified*, we know that most designers compare different possibilities, exploring why one might be better (or seeking ways to combine the best of both).[13] Another term for this is a tradeoff: finding the balance between two (or more) competing goals. For example, in visual design, the concept of composition is about making tradeoffs between whitespace, hierarchy, and contrast to find balance. If you don't do this, and everything gets equal attention, it's probably a confusing, ugly failure. Similarly, a product that tries to do too many things, like a Swiss Army Knife, probably doesn't do any of them particularly well. Making tradeoffs is central to what designers do.

A great lesson about tradeoffs is found in an old adage, often seen on signs in dive bars and diners: "We offer three kinds of service: GOOD, CHEAP & FAST. You can pick any two." It sounds like a joke, but this same idea in the business world has been called the **iron triangle** since the 1950s.[14] It's a real law of constraints that experienced project leaders know well because it captures how hard it is to make quality things. The triangle is a reminder to good leaders that they must make tradeoffs to succeed, and that's often done by setting clear, realistic goals. On the other (much stupider) hand, bad leaders assume they are immune from the triangle (i.e., the Dunning-Kruger effect), which leads to failed projects that are slow, expensive, and bad, the trifecta of failure (aka the failfecta).

---

13 Bryan Lawson, *How Designers Think: The Design Process Demystified*, 4th ed. (Oxfordshire, UK: Routledge, 2005).
14 "Project management triangle," Wikipedia, https://en.wikipedia.org/wiki/Project_management_triangle.

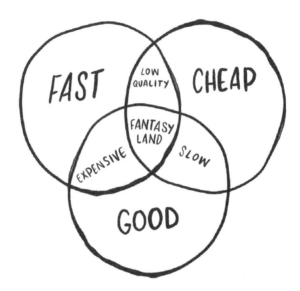

For designers, the breakthrough is to use our tradeoff talents when working with other people. Being good at making tradeoffs with others is called negotiation. Good negotiators do more than just fight for their own position; they're able to compare their own interests with the interests of the other side. They're curious about the real reasons the other party has its position, rather than judging them for it. If we use our curiosity, we can uncover flawed assumptions in the other party's thinking, or our own, leading to happier outcomes for everyone.

However, the ego-trap view is that negotiation must be an "us vs. them" battle. This attitude makes us defensive and prone to pre-emptive strikes to protect our ideas, which works against us as advisors. Design culture, and the world at large, still treats negotiation and persuasion as secondary abilities (i.e., the pejorative of "soft skills"), but in practice they are primary ones, especially for people in advisory roles. So much of our time is spent negotiating with leaders, engineers, marketers, and clients. If we choose to treat negotiation as a central role we can play, we become far more influential with deciders.

The book *Getting to Yes*, by Roger Fisher, William Ury, and Bruce Patton, should be required reading in design school because it provides a simple way to reframe what we see as adversarial situations into creative thinking and relationship building. School design projects should involve relationships as much as they involve ideas.

Whenever there's a proposed decision you disagree with, before you get upset, investigate. Consider these questions:

- What tradeoff is being made?
- Is the decider aware of the trade (i.e., the iron triangle)?
- Do you and they have the same goals? How can you align?
- Who, if anyone, is arguing the side of the tradeoff you prefer?
- What investigation or explanation can make this position stronger?
- If no one is arguing for it, why do you expect it to win? Who should do it? Why aren't they?

Every decision can be framed as a possible negotiation. It's just a matter of who takes the initiative (and if this scares you, we explore introversion and extroversion in Part 3). Design becomes easier when you become a confident negotiator—then you won't be reliant on other people to do it for you.

Reframe Your Habits

Abraham Maslow, of "Hierarchy of Needs" fame, once said, "If the only tool you have is a hammer, you tend to see every problem as a nail."[15] Ironically, the Hierarchy of Needs theory itself is used in countless ways

---

15 "Law of the instrument," Wikipedia, https://en.wikipedia.org/wiki/Law_of_the_instrument.

he didn't condone, but that's a story you can investigate on your own.[16] As designers, we have many hammers we use too often and it gets us into trouble. As Bryan Lawson wrote in *How Designers Think*, "One of the most important skills designers must acquire is the ability to critically evaluate their own self-imposed constraints."[17]

We have habits that are strengths, but when taken too far, they work against us. To make our jobs easier, we need to step back and reevaluate our tendencies.

Output Worship

We love making things. We make prototypes, screens, diagrams, processes, taxonomies, and more. Many of our hobbies involve making things, like photography, music, or even furniture. Our most central belief might be that creating things solves problems (or at least distracts us away from problems we can't solve).

Where we run into trouble is when the real problem is not solved by creating more things. What good is creating another idea engineers will ignore? Or proposing another team process that no one will use? We're prone to the false belief that if we can make something smart and beautiful enough, all of our problems will fade away. It's a fantasy, born from TV shows like *Mad Men*, that when powerful people, who historically have low design knowledge, see how "brilliant" our idea is, they will instantly drop all of their past resistance and the sunk cost of their current idea and follow

---

16 Gareth Harvey, "Why Everything You Know About Maslow's Hierarchy of Needs Is Wrong," LinkedIn, June 13, 2023, https://www.linkedin.com/pulse/. why-everything-you-know-maslows-hierarchy-needs-wrong-gareth-harvey/
17 Lawson, *How Designers Think*.

our lead. This output worship is a story only someone in the ego trap could believe. And in the ego trap, when this approach inevitably fails, we blame ourselves for not being brilliant enough, or our leaders for not seeing how brilliant we are.

There is a better explanation. Design educator John Spencer said:

> The most pervasive stereotype about designers is that our job is all about doing things that are inventive, ingenious, and inspired. And that design has to be "creative" to be any good. It's a stereotype that's perpetuated by designers—and it's bollocks. Creativity is a distraction. More than anything, our job is to make sense of things. It isn't about being clever—it's about being clear.[18]

To put this another way, it's dangerous to confuse outputs with outcomes. An output is just something that was created, but it may not lead to the desired outcome. If we want the outcome to be leaders using our ideas, it requires us to understand their thinking and to gain their trust. Instead of creating volumes of output, we just need to have a clearer, stronger idea of what they want, based on their criteria, and deliver that solution.

Of course, sometimes it's effective to wow people with our creativity and to intentionally try to change how they think. Sometimes deciders need to see something before they will believe it. We're just saying do this with your eyes open: focus on the outcome more than the output. When we focus on outcomes, there's usually a less stressful way to get what we want.

---

18 Sarah Dawood, "What Do You Think Is the Most Common Stereotype About Designers?", *Design Week*, February 19, 2015, https://www.designweek.co.uk/issues/february-2015/what-do-you-think-is-the-most-common-stereotype-about-designers/.

Elitism

Designers love to feel special, and there's nothing wrong with that. We are special. Who else organizes their bookshelf by the color of the book covers? The tradeoff is that if we want people to include us and treat us as equals, being special complicates things.

As a professional minority, we often prefer to be with other designers, despite the reality that the deciders we wish to influence are not part of this group. If we're not careful, we reinforce habits that make design harder. Instead of relying on our skills of investigation and explanation, we become entrenched in defending our vocabulary and methods. The ego trap is tempting because we can always find people like us, with the same complaints and preferences, but it can be a comradery of disappointment. The people in it are too disillusioned, or too afraid, to use their open minds to reframe problems into actions.

Paul Adams, Chief Product Officer and former designer, realized that he needed to grow out of the trap, as he expressed in his classic talk, "The End of Navel Gazing":

> I was naïve, I was biased, I was ignorant, and I was prejudiced. I was all of those things about all those other disciplines, and I never really thought too much about it...because I was too busy having existential crises with my colleagues about what the hell we even do, and why no one listens to us.[19]

Hodges-Schell and O'Brien describe what many coworkers say when we're not around:

---

19 Paul Adams, "The End of Navel Gazing," *Intercom*, https://www.intercom.com/blog/the-end-of-navel-gazing/.

> To those outside traditional design roles, design and designers can appear intimidating, unquantifiable, and having a different set of rules… Lack of trust is the ultimate anti-pattern, but by involving stakeholders and demystifying our processes, it is possible to fix this. Trust runs both ways, of course. Why should an organization full of stakeholders trust design if they feel as if the designer doesn't trust them?[20]

What would happen if your team saw you as helpful? Insightful? Trustworthy? Useful? We bet they'd invite you to more important meetings than if you were combative, distant, or cynical. Elitism is only an advantage if you have power, and even then it's a dubious way to treat people. Rams brought brandy to his conflicted team. What do you bring? If you're truly an expert on human nature, apply those basic lessons of charm and hospitality to your coworkers. We're not saying do unrewarded labor or be taken advantage of. We don't want that for you at all. Instead, we're asking you to question your habits and to investigate if they are serving your goals.

A great way to grow is to become curious about the expertise you don't have. Instead of taking more design courses, where you're already an expert, get out of your comfort zone and learn from people who aren't designers. Don't sign up for the "MBA for Designers" class assuming it's better optimized for you. In a way, the opposite is true because it keeps you in the design bubble, which you need to get out of. Make it a goal to befriend, and learn to empathize with, people from other jobs and backgrounds (which job roles annoy you the most? Engineers? Business Analysts? A bold move is to start there). Taking a course or attending non-design events lets you meet

---

20 Hodges-Schell and O'Brien, *Communicating the UX Vision*.

engineers, marketers, and entrepreneurs as peers. As your understanding of how they think improves, you'll find cross-functional gravity problems get easier.

Detail Obsessions

We know how much pixel perfection matters. For many of us, it's the pleasure (or compulsion) to control every little detail that led us into this profession. The risk is that this makes us prone to obsessing about trivial matters when much larger problems must be solved first. In project management jargon, this is called a **scope mismatch**. An example is when a leader is trying to get consensus on overall project goals, but an engineer or a designer (both are prone to detail obsessions) insists on first discussing how to implement a specific low-level potential feature that may not even be needed.

When we insist on too many details too early, it frustrates deciders. It makes them feel like we don't belong in the conversation, since we're focusing on the wrong scope of the problem.

Obsessions we indulge in often include:

- Labels: Names for features, products, methodologies, job titles
- Railroading: Resistance to collaboration, insisting on one single path
- Preciousness: Sensitivity to feedback and an unwillingness to discard ideas
- Refinement: Demanding precision too early in the process

A related obsession we suffer from is fear of showing low-fidelity, or quickly sketched, work. The fear comes from being unfairly judged for not having all the details figured out, or perhaps even PTSD from prior abusive design critiques. We forget we're rarely working with other designers. Our laziest sketches are better than what many of our business and engineering coworkers could ever produce, and they will be impressed by them (a talent mismatch we should exploit). It's also good to know that people give feedback on the level of fidelity provided, and for early project questions, low-fidelity work (i.e., hand-drawn or black-and-white) leads conversations to the most productive place.

Another flavor of detail obsession is what is called **bikeshedding**, or Parkinson's Law of Triviality.[21] The term comes from a committee created to approve a nuclear power plant, but instead of discussing important matters (you know, like nuclear fallout), they argued for hours about the bicycle storage shed. Why? Because it was a topic simple enough that everyone felt qualified to have an opinion (the Dunning-Kruger effect strikes again), whereas the important decisions were complex and required deep knowledge that few people had. Bikeshedding is a kind of theater, a show that allows people to believe they're adding value when, in reality, it's just a distraction.

Design job titles and methodologies are classic examples of bikeshedding. There is, and will always be, endlessly dumb arguing about what design is, what designers should or shouldn't do, and who is or isn't a designer (some will say us authors do not qualify). Why is it dumb? Because it's usually philosophizing masquerading as work. It often has zero effect on gravity problems because the deciders we wish to become, or to influence, do not care about bikeshedding topics. They are too busy making real decisions on actual projects in the world.

---

21   "Law of triviality," Wikipedia, https://en.wikipedia.org/wiki/Law_of_triviality.

The people who spend the most time bikeshedding are often the most disenfranchised in their organizations, or are consultants seeking attention to help find clients. They might truly believe that "if only everyone would agree on my definition of <*insert word that has plenty of good definitions already*>, gravity problems would disappear," but this is another ego trap. It perpetuates the perception of designers as elitists who do not understand why design is hard and who are useless in helping get real work done.

Factionalism

In *Monty Python's Life of Brian*, an absurdist comedy about the early Roman Empire, multiple rebel groups share the same goal of overthrowing the Romans. When Brian meets one group for the first time, he asks, "Are you the Judean People's Front?" They respond in shock and anger that he would suggest this, as they hate the Judean People's Front. They yell at him, as if it should be obvious, "We're not the Judean People's Front. We're the People's Front of Judea!"[22] This is a tendency known as the **narcissism of small differences**: we tend to fight most with the people we have the most in common with, often over trivial details.[23] It's a pattern you can find in professional circles but also in ethnic conflicts and civil wars.

The habit of bikeshedding is a leading cause of factionalism, where groups who share the same larger goal (a better-designed world) end up in trivial debates and rivalries. Design is a small enough professional minority to begin with, but when graphic

---

22 *Monty Python's Life of Brian*, directed by Terry Jones (Cinema International Corporation, 1979).
23 "Narcissism of small differences," Wikipedia, https://en.wikipedia.org/wiki/Narcissism_of_small_differences.

designers, UX designers, content designers, service designers, design researchers, and whatever new sub-classifications of design we come up with next focus on our differences more than our similarities, it fuels the ego trap. Instead of forming a community around our common need to overcome the gravity problems that affect us all, we indulge our detail obsessions instead. Specialization is valuable, but if we want design to become easier, we have to help each other solve shared situations that make design hard.

PART 3

Why Design Is Hard

WELCOME TO THE halfway point of this book. Since you're this far along, you might be wondering why this part shares the same title as the book itself. It's almost as if we're suggesting that the hard things we've covered so far were just the preliminaries and it's about to get harder. If that's what you're thinking…you're correct! We'd like to give you a prize here, but it would have to be the least popular kind there is—what we call a truth prize.

A truth prize is really just a nice way of saying that you have more problems to deal with than you thought. A good example is when your toaster isn't working, and after spending hours trying to repair it, you finally realize that the outlet it's plugged into doesn't work. This is not much of a prize really, but we warned you it's all we have. We think it's good to get truth prizes because the alternative, continuing to solve the wrong problem forever, is worse. We believe in the African proverb that "a smooth sea never made a skilled sailor." And since you've made it this far without throwing the book at the wall, or at least if you did, you dusted it off and picked it up afterwards, we know you're ready for rougher waters.

Bad Design Makes Money

One question designers hate to answer is: why are we here? Why should companies hire designers? The common answers we offer, like "we fight for the users" or "we make products better" come from design idealism. We resist the painful truth that companies hire us because they expect us to generate more revenue than we cost. That's it. Many companies don't hire designers. Why? They don't expect them to generate enough profit or to earn what they'd have to pay us. It's not complicated. We're not special in this regard: all employees are evaluated in the same cold way.

Many designers are shocked to learn that:

- **Bad design is profitable.** Better design means improving quality. Quality is expensive, and expenses eat into profits. Look at the market for any product, like cars or mobile apps: what percent are designed well? 20%? 5%? Many "lesser" products we look down upon are made by profitable companies. Put simply, affordability often comes before quality in driving sales, meaning that many businesses can't afford the cost of high design quality. Some frustrated designers think they're Michelin star chefs and are in denial that they're employed by the equivalent of the local pub.
- **Business is often just math.** Let's say MegaStinkCo has 30% market share with its MegaStink7 product. Its leaders want to get to 40% but aren't sure how. They could invest in design

("stinkovation!"), but leaders have little experience with that strategy, so established avenues like marketing, advertising, and partnerships seem wiser. It comes down to math and predictions: if leaders guess that design as a strategy costs 10X to reach their goal, but marketing costs 3X, which should they choose?

- **A monopoly with a weak product can profit more than a strong product without one.** With zero competition, product quality is irrelevant: a company will dominate the market (and maximize profits by not investing in raising quality). For this reason, many businesses seek platform dominance because it grants them effective monopolies. Tech giants, like Microsoft, Oracle, and Adobe, have invested billions to make their platforms dominant, cultivating customer dependency, so that switching to competitors with better products is too cost and time prohibitive.

Of course good design is typically good for business. It's just not as universal as we're led to believe. When the goal is one of the common ones of gaining market share, improving customer retention, or developing a new product, design can be a strong strategy. However, many businesses choose other strategies, like price or promotion. This means **design for sale** is the priority: making choices that help drive sales, like adding more features, even if it makes the product harder to use.[1] For example, microwaves and other appliances often have terrible user experiences because having more features than competitors is a classic design for sale tactic that works. Advertising is a common design for sale strategy in commodified products, like soft drinks, automobiles, and fast food, where billions are spent

---

1 Victor Papanek, *Design for the Real World: Human Ecology and Social Change* (Chicago: Chicago Review Press, 2005).

hiring very creative people to convey an (exciting) experience before the (bland) product is ever purchased.

This is very different from **design for use**, which means making the best user experience. Most designers erroneously assume that design for use is the most important business goal. Good counter examples include subscription businesses like your local gym or a dating app. These models profit most from customers who sign up and pay, attracted by sales and advertising, but rarely show up to use the services. Once they're subscribers, they're all profit and no cost. And if the gym is the only one for 50 miles, they also have a monopoly, which means improving the gym itself might not increase revenue.

As design strategist and consultant Erika Hall explained, we've cultivated the false belief that what is good for customers must always be good for business. It's an assumption we rarely question:

> Highly-financialized businesses, platforms...and money-losing speculative nonsense alike, don't operate according to the straightforward "more user value = more business value" equation that has been treated like a law of physics in design.[2]

In our ignorance, designers often fight for ideas, unaware that they work against the business goals of our employers. And instead of investigating the business logic that explains the resistance we experience, perhaps finding the sweet spot where design for sale and design for use overlap, we resort to moral arguments ("we *should* do the right thing"), which we rarely win. This doesn't mean giving up or giving in.

---

2   Erika Hall, LinkedIn, https://www.linkedin.com/posts/erikahall_once-again-i-am-seeing-people-blame-designers-activity-7156370314241736704-XuKx/.

Of course we should speak up to prevent bad things from happening in the name of profit, but we need to be smart about it.

We should stay dedicated to the goal of producing quality things and making a better society for everyone. As W. E. B. Du Bois said:

> We should measure the prosperity of a nation not by the number of millionaires but by the absence of poverty, the prevalence of health, [and] the efficiency of public schools.[3]

However, we need to be wise about how, when, and where we pursue this. For example, if we're hired by a megacorporation like MegaStinkCo to improve its products, we must remember this rule from designer and author Julie Zhuo:

> If your company isn't profitable after some period of time, everyone loses their jobs.[4]

Most corporations are designed, documented in their own by-laws, to extract wealth from any source and give it to shareholders, not customers, not employees, and not society. It's foolish to pretend we work for altruistic organizations whose leaders have society's best interests in mind. They don't. But to Zhuo's point, if we can make arguments based on which position is most likely to help our company win, rather than ones about doing the right thing, the profit incentive works in our favor. Hodges-Schell and O'Brien add that "to persuade people that our solution is right, we must first convince them that it isn't contrary to their definition of right."[5]

---

3   W. E. B. Du Bois, "On the Future of the American Negro," 1953.
4   Julie Zhuo, "The Looking Glass: What Company Politics Actually Is," Substack, October 24, 2023, https://lg.substack.com/p/the-looking-glass-what-company-politics.
5   Hodges-Schell and O'Brien, *Communicating the UX Vision*.

Erika Hall wrote that "design is only as humane as the business model allows and rewards."[6] But there's more to the story. In the U.S., we once had healthier checks and balances between business and society, but the government's ability to do this has been stripped away since the 1970s through short-sighted deregulation. Capitalism does have advantages, but only when our collective long-term interests are protected (e.g., compare the causes of the financial crash of 2007 with the philosophy of W. E. B. Du Bois). The European Union does this better in some ways, which is why food in Europe is healthier and tastes better, and the biggest anti-trust challenges to billionaires and megacorporations come from the EU and not America.[7]

But instead of feeling hopeless, remember that design, as a skill, predates these challenges by millennia. Our ability to make tools and solve problems collaboratively is central to why our species has survived for the last 300,000 years.[8] As a designer, your abilities are the embodiment of what makes humans special. So if you have the choice, consider working for people who share higher values than profit above all. If you can use your talents in service to future generations, rather than just who pays the most, you can become part of the solution—designing a better future for everyone.

---

6   Amy Lee, "Clarity Recap: 'The Business Model Is the Grid,' Erika Hall," Medium, November 6, 2023, https://medium.com/@amster/clarity-recap-the-business-model-is-the-grid-erika-hall-2f53f31abe40.
7   "Food Safety Regulations in the European Union," Biosafe, https://www.biosafe.fi/insight/food-safety-regulations.
8   Brian Handwerk, "An Evolutionary Timeline of Homo Sapiens," *Smithsonian Magazine*, February 2, 2021, https://www.smithsonianmag.com/science-nature/essential-timeline-understanding-evolution-homo-sapiens-180976807/.

Systems Come First

Imagine what would happen if we had a time travel machine and transported Dieter Rams, in his prime, to a job at a small farm on the outskirts of Cairo, Egypt, in 1874. How successful would his product design career have been without indoor plumbing or electricity? In his book *Out of the Crisis*, author W. Edwards Deming estimates that 94% of the success of an organization is its systems, rather than the behavior of individuals. That's a shockingly big, and curiously specific, number to hear at first. But think about how many necessary factors there are for a project to succeed that are greater than what any one person contributes. From that point of view, Deming's estimation makes great sense.[9]

Consider that we don't pick the year we are born, what country we are born into, or who our parents are, yet these are likely the three most important factors that define our lives. The individual choices we worry about pale in comparison to the systemic factors we never had control over in the first place.

In her classic book *Thinking in Systems: A Primer*, Donella H. Meadows offers this definition:

> A system is a set of things—people, cells, molecules… interconnected in such a way that they produce their own pattern of behavior over time.[10]

For our purposes, the most interesting thing about systems is their interconnectedness. It means that each part has many relationships to other parts, and the system works (or fails) for a chain

---

9 W. Edwards Deming, *Out of the Crisis* (Cambridge, MA: MIT Press, 1982).
10 Donella H. Meadows, *Thinking in Systems: A Primer* (White River Junction, VT: Chelsea Green Publishing, 2008).

of reasons, rather than because of a singular cause. For example, when only one tire on your car wears out, you might just think it's a bad tire and replace it. But the real cause might be that the tires are out of alignment, or there could be a problem with the suspension system and not the tire itself. Humans are prone to shallow thinking and we're attracted, like a moth to the flame, to simple explanations. Explanations based on singular causes are easier to say and remember, even when they're wrong. Thinking in systems usually provides better explanations for why things are how they are.

Think about the last time someone at your workplace did a great job on a project or made a terrible mistake. Odds are they got much of the credit or the blame. But if we think in terms of systems, we can ask better questions that help us really understand what happened:

- Who decided what work this person was assigned?
- How was this person hired, trained, and managed?
- How were they rewarded, supported, or punished?
- What powerful people enabled this situation or failed to prevent it?
- Did this event happen in line with the team culture or in rejection of it?

Just a handful of questions improve the quality of our explanation. Similar techniques like root cause analysis or the 5 Whys (i.e., ask why repeatedly to understand something's true cause) serve the same purpose. Using these tactics when you're frustrated or stuck can help widen your perspective on what's really going on.[11] Renowned scholar Theodore Zeldin wrote, "The people with the most power…are those that have been trained to look at the world

---

11 "Five whys," Wikipedia, https://en.wikipedia.org/wiki/Five_whys.

as a series of systems, not of individuals, and who therefore can produce systematic solutions." We'd add that this ability to see in terms of systems is why some people are able to obtain power, despite not being born into it or having it gifted to them.

When we look at our workplaces in terms of systems, it's easier to see that the friction we experience is probably an efficiency for someone else, usually someone more powerful than we are. Because of our output worship and detail-obsessive tendencies, we're prone to waiting pridefully for organizations to conform to the way we want to work (e.g., throwing design work over the fence to engineering). Even if we don't want to do the work to close the gaps we find, we have to be mature enough to see that if the gap remains, it works against us eventually.

This realization often leads to another flawed approach designers are fond of: revolution! We try to take on a whole system ourselves. It feels heroic, and movies and myths make it seem realistic, but when you work against a system, it's like swimming upstream in a powerful river. You can be talented and hard-working, but even simple actions are hard. You will exhaust yourself. Even if you have some success, the inevitable burnout might take years to recover from. We don't recommend doing this. The place you work for is unlikely to be worthy of this kind of sacrifice.

Instead, step back and, this will sound strange at first, empathize with the system. If you think of your organization, or the river in our metaphor, as an organism, it's just doing its job to survive and thrive (or to externalize costs and maximize profits). The river pushes things one way because that's what it has been doing successfully for a long time. Even inert elements, like leaves and branches (or bad middle managers), help the system too, because once they fall into the river and float along with it, they work against anything trying to go the other way. Successful systems are self-reinforcing, which makes them powerful.

So what does this mean? You don't have the luxury of assuming the system you work in is designed to make the high-quality design work you want to do easy. It almost certainly won't be. Instead, use your investigative powers to:

1. Make good maps to ease navigating the system
2. Look for points of leverage where your assets are greater than your liabilities
3. Find allies in the system so their status becomes an advantage

Because these aren't design tasks in the strictest sense, your idealistic mind might resist and ask our familiar question, "Why should I have to?" And your creative brain might be thinking instead, "Why can't I redesign the entire system from scratch and make it better?" Even if you had the power to do this, which you don't, you'd likely fall victim to the **blue-sky fallacy**. Creative people frequently fantasize about having a clean slate and starting over (i.e., the only constraint is the blue sky above us). Most of our design methodologies suffer from blue-sky thinking: they somehow assume a blank slate of a team waiting for a method, instead of realizing there's a pre-existing power structure that leaders like and have no interest in changing.

Part of the blue-sky fallacy is the lack of awareness that every single system that disappoints us with its brokenness and sheer stupidity started as something that someone somewhere designed from scratch. It almost certainly worked well when it was first made, but over time, unanticipated challenges accumulated and the system declined. It's only ego and ignorance that make us think the solution is to give one person—the designer, of course—supreme power to start everything over. As tempting as it is, this is not the way.

Instead, the path to redesigning systems begins with what is known as **Chesterton's Fence**:

> You should never destroy a fence, change a rule, or do away with a tradition until you understand why it's there in the first place.[12]

By choosing to be an investigator first, you'll actually understand the problem you really need to solve. For example, often designers discover that their organization is siloed, or overly divided, which makes good UX design harder. Don't assume everyone wants to change this. Instead, find out who benefits from the status quo (e.g., busting silos shrinks the empire of selfish silo leaders). Then compare that list with who, besides you, benefits from a shift to a more holistic culture.

For organizations, good investigation begins with making your own useful maps. While existing ones like org charts and official processes tell one story, and probably not an accurate one, bet on your curiosity instead. In their book *Hack Your Bureaucracy*, Marina Nitze and Nick Sinai explained:

> Relationships, knowledge, and power aren't captured by traditional org charts. [Make] your own map of who knows what, who controls what resources, and who is in cahoots with whom.[13]

---

12 Jonas Koblin, "Chesterton Fence: Don't Destroy What You Don't Understand!", Sprouts, June 1, 2023, https://sproutsschools.com/chesterton-fence-dont-destroy-what-you-dont-understand.

13 Marina Nitze and Nick Sinai, *Hack Your Bureaucracy: Get Things Done No Matter What Your Role on Any Team* (New York: Hachette Go, 2022).

Never underestimate the power of cahoots. Organizational systems are made of people, and people behave in preferential ways to people they like. Leaders often feel the need to obscure their preferences under the guise of professionalism, trying to give their behavior the illusion of Vulcan-like rationality. When you can see past it, it may explain some of the behavior that has mystified you in the past.

To construct your map, consider the following questions:

- What is the real path ideas follow to get approved?
- Who are the true influential players on that path?
- What secret passages or backchannels do they use?
- Who trusts you? Likes you? Avoids you?
- Who owes you a favor?
- How diverse, in role and background, are the people you talk to every day?
- Which powerful people are looking out for you, or could be?

With good system maps, you will find **leverage points**. Meadows defines leverage points as "places within a complex system where a small shift in one thing can produce big changes in everything."[14] For designers in organizations, they often include things like:

- A person at the right level of power who trusts you
- A project is small enough that you can successfully influence it
- The team, or its leader, is open to a new approach
- An ally with inside knowledge to guide you in selling your ideas
- You have experience with the goal that the team wants to achieve

---

14 Meadows, *Thinking in Systems*.

When you have leverage opportunities, you'll no longer feel like you're swimming against the current of the river. You'll see the breakpoints where the current changes, sunny banks providing places to rest, and even (solar-powered) motorboats your allies are driving that have space for you, taking you in the direction you want to go. You don't need the whole river to change all at once, you just need to invest where the system works in your favor, so you can grow more leverage points and repeat.

Power Goes to Generalists

Think about your boss and how their job is different from yours. Now go up two more levels and ask the same question. In almost every organization, the higher up you go in an org chart, the more generalized, and powerful, the roles become. The CEO in this sense is the ultimate generalist: they have the most power, the widest responsibility, and the highest pay. It's the natural order of how most organizations work: there is a hierarchy of power that defines the culture. The reason you are more of an advisor than a decider is likely because you are more of a specialist than a generalist and lower in the hierarchy.

One obvious way to resolve this, that designers resist, is to move into generalist roles. If a generalist, say a project director or even a product manager, makes the important design decisions you wish were yours, then why don't you just take one of those jobs?

Designers usually resist this idea because we chose this career so we could do creative work, and we have little interest in the multitude of non-creative-seeming tasks that generalists have to do. And this is OK. There's nothing wrong with your preferences for what you do all day, or for wanting to specialize in, as opposed to generalize, the work you do. The tradeoff is that in any system, your preferences may not come with the power and pay you desire. Leaders design organizations to succeed as a business, not primarily to make employees happy.

Specialization has other risks too. It can make it harder to see the larger goals we claim to care about. In the book *Culture Care: Reconnecting with Beauty for Our Common Life*, artist Makoto Fujimura explained:

> A related cultural fault line is hyperspecialization, where a person or firm focuses on increasingly narrow segments of a production process, a discipline, an artistic genre, or a market…the expert knows one part, not the whole, and often not even the wider field in which they work. They consciously reduce their scope of concern to go deeper in their discipline. But increased clarity on a narrow point usually comes at the price of blindness to context and to one's working assumptions. It often brings isolation from—and sometimes alienation from or hostility to—those with differing expertise.[15]

Designers often feel tension with the nearest powerful generalist. This might be product managers and project managers, but just as often those roles feel underpowered and marginalized too. If you're not sure, ask them. They may share your feelings and can become allies. It's not uncommon for the real generalist who decides everything to be at the director or executive level. They might just be one of those people who doesn't realize how much micromanaging they're doing, or who doesn't care that many talented people are frustrated by their destructive habits.

---

15 Makoto Fujimura, *Culture Care: Reconnecting with Beauty for Our Common Life* (Westmont, IL: IVP, 2017).

It helps to remember that, in most cases, narcissistic executives aside, it's not the generalist's fault that there is tension with designers. Some of that tension is healthy, if it's managed well. Jerry Hirshberg, the former design director for Nissan, calls it **creative abrasion**.[16] Healthy friction and debate can bring out the best ideas from everyone. It's through a diversity of opinions and backgrounds that the best work often happens.

More broadly, all organizations need specialized and generalized roles to function. If you assembled a superstar team of the world's best specialists, you'd likely fail to make anything good unless you had someone skilled at leading, negotiating, planning, communicating, and managing up, and who was also entrusted to make tradeoffs between specializations. This is what film directors, who simultaneously play the role of artist and entrepreneur (in partnership with producers), do as the core of their jobs.

Another reason designers hate the idea of moving into generalized roles is because that drops the word *design* from their job titles, which feels like the loss of their identity. But this might just be a detail obsession. If you had to make the tradeoff, is it better to be a designer in name or a designer in reality? We think your fulfillment as a creative person will come primarily from putting good things into the world, rather than what your business card says or which specific kinds of tasks make up your day.

Let's imagine you work at MegaStinkCo as the lead product designer. It's a good company, filled with thoughtful, intelligent, and collaborative people. The problem is that you often feel ignored because

---

16   Scott Berkun, "How Creative Friction Can Help Your Team," *Scott Berkun* (blog), May 8, 2018, https://scottberkun.com/2018/how-creative-abrasion-helps-good-ideas/.

others make the decisions. You sometimes think the system should be restructured or that designers should be trusted more, but you don't have the power to win *should* arguments. If we diagrammed it, it would look like this:

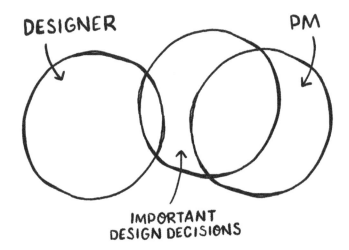

Treating this situation like a gravity problem, there are only three choices. In Part 1, we said the second lesson of this book was to accept that you must either seek power, gain influence, or become more self-aware. Applying that lesson here gives these possibilities:

A. **Choose organizational power:** Take the role most aligned with the decisions you want to make.
B. **Choose influence:** Stay in your role, but earn trusted allies and negotiate power.
C. **Accept gravity:** If you don't want to change (A or B), that's OK, but blame others less.

Blaming others, which includes making empty should arguments, is to fall back into the ego trap. Choices A and B, on the other hand, have a good chance of making you more fulfilled, because the work you want to do is better aligned with the role. As a decider (A), coworkers will learn more from you about good design, and at a faster pace, since you're in the middle of things. For choice B, as you earn trust, you can negotiate to have more power to make decisions. With option A or B, it won't take long before the diagram looks more like this:

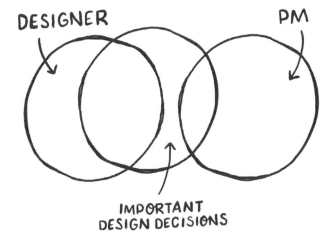

Neat little diagrams in books make it seem so easy, but we know it isn't. Sliding a shaded circle around in Adobe Illustrator takes seconds, but choice A or B could take weeks, months, or years to accomplish, and since it's likely outside your comfort zone, there will be some growing pains and risks. That's all true.

A larger point worth noting is that negotiating roles and redistributing power is a universally wicked problem. Right now in your organization, executives and team leaders are in meetings debating who should get to decide all sorts of things. The U.S. Senate and the United Nations debate power and authority all day long. If your role frustrations make you upset, it might help to realize that you're experiencing a taste of the challenges taking place everywhere important decisions are made. Healthier organizations make this less painful through trust, communication, and good leadership, but it's still hard work, and rarely is everyone happy.

Part of the problem is the fantasy of having the fun parts of power but none of the downsides. This does not exist. Power comes with unpleasantness no one wants, like higher stress, dealing with difficult people, and being responsible for things you can't control, including ugly, messy hot-potato problems everyone else was smart enough to avoid. The upsides—deep satisfaction, joyous collaboration, and getting paid to make your ideas real—outweigh the downsides for many people (reminiscent of the stories about Paula Scher and Dieter Rams in Part 1).

If you don't want organizational power, remember why you made that choice when you're trying to influence someone who has it. Their job is harder than it seems. What can seem like stupidity is often a constraint you can't see, like a compromise between people even more powerful who don't get along. As UX leader Doon Malekzadeh told us, "There is always one piece of information that can radically change your understanding of a situation." When you're frustrated by your organization, investigate the real causes.

Regardless of what role you have or want, you can do simple things to grow your generalized influence, including:

- **Taking notes and reporting.** Don't be deceived into thinking this is menial work. The scribe for meetings creates the history. If you're new to a team, it's a low-risk way to earn trust and gain visibility. When people see that you thoughtfully capture what they heard and said, and that you explain things clearly—perhaps through visuals, which no one else on the team can do—you gain credibility that will help later.
- **Facilitating discussions.** In Part 2, we talked about the power of explaining and negotiating. If you help people understand each other, often by leveraging your design skills, people will want you in the room for important conversations. Good

communication is the basis for influence. If you're frustrated by the low quality of feedback you get on your designs, you can get better feedback by becoming a better facilitator.
- **Clearing paths and investigating important problems.** The person who clears the way shapes the path. If you want change, it's more likely to happen if you lead rather than follow. Good leadership is rare, and when people see it done well, they will follow along.

Organizations Reward Connectors

For once, let's imagine a scenario where everything goes right:

> You are the lead designer at MegaStinkCo working on a blue-sky project to redesign the flagship product (MegaStink8). Everyone does great work, on time and under budget. Designers, PMs, and engineers get along smashingly well, dancing in the hallways and singing on Zoom calls. You balance design for sale with design for use, doubling revenue and tripling all UX metrics, while reducing corporate greenhouse gas emissions by 420% and creating jobs for 4,812 deserving, unemployed, unhoused citizens. The Universe is so inspired by the brilliance of your work that dogs and cats become friends, all wars end, racism disappears, and hell freezes over. Your VP asks for someone to present the project results at the next executive briefing. It's a five-minute slot so only one person can present. Who should do it?

No matter who gets chosen, that person will become more visible. More leaders will know her name and be more likely to respond to her requests. She will get more questions about the project than others in the future, and her opinions will carry more weight. It's not anyone's fault, it's the simple fact that visibility has advantages. It's harder to trust someone you've never met or heard from. Many people shun visibility, or label themselves introverts, avoiding public speaking or exposure of any kind. Assuming they are good at what they do, this almost certainly works against them.

We're not here to transform you into a mega extrovert. That's not necessary and would likely make you very annoying to your coworkers. However, we do want you to know three things:

1. **Introvert vs. extrovert is a false dichotomy.** Most people are on the spectrum between extremes. In one study, 77% of people reported being somewhere in the middle.[17]
2. **The best definition of these terms is about sources of energy.** Introverts tend to want to be alone to recharge, whereas extroverts tend to get energy from others.
You can be an introvert with excellent social skills, or an extrovert with terrible ones (which may help you better understand certain people in your life).
3. **These are not genetically encoded traits that never change.** Your desire for social interaction almost certainly shifts depending on mood, context, environment, and other factors.

These facts are important because organizations are systems of people. The more open and curious you are about people, the more the system is an advantage. If you go out of your way to make

---

[17] Jennifer Guttman, "Introvert vs. Extrovert: How Does It Affect Social Anxiety?," *Psychology Today*, February 15, 2021, https://www.psychologytoday.com/us/blog/sustainable-life-satisfaction/202102/introvert-vs-extrovert-how-does-it-affect-social-anxiety.

connections, the system will provide more useful maps and leverage points. Alternatively, if you go out of your way to avoid connection, you will have fewer. Maybe you don't need many and have all the power and influence you need—flying under the radar can work sometimes. But if you're reading this book, odds are that's not the case.

Being a connector in an organization is a natural role for designers, because it's at the heart of all the work we do. Most kinds of design are about communication, and the powers we explored in Part 2 included investigation, explanation, and negotiation, which all make connections between ideas and people. The more you can use your abilities to be a connector, the easier design work becomes.

A powerful metaphor that helps is what Makoto Fujimura calls **border-walking**. He describes this as people who choose to spend time on the edges of ideas, cultures, and tribes. He explained that:

> These were individuals who lived on the edges of their groups, going in and out of them, sometimes bringing back news to the tribe....[They] lower barriers to understanding and communication, and start to defuse the culture wars....[It's] a role of cultural leadership in a new mode, serving functions including empathy, memory, warning, guidance, mediation, and reconciliation. Those who journey to the borders of their group and beyond will encounter new vistas and knowledge that can enrich the group.[18]

Does this require being an extrovert? Not at all. Instead, it requires being curious and taking initiative, which anyone can do. Most people appreciate those who:

---

18   Fujimura, *Culture Care*.

- Are genuinely interested in the work they are doing
- Ask good, thoughtful questions
- Have respect for their skills and beliefs
- Are helpful

If once a week you "walk the border" between teams, roles, hierarchies, or wherever there's the most tension or ignorance in your world, and embody the list above, you will be making connections. Good leaders do this frequently, sometimes calling it **management by walking around (MBWA)**, or in the parlance of Toyota's Lean philosophy, a **Gemba walk**, the act of seeing where the work happens. Rams's legendary bottle of Cognac was inspired by his border-walk into the engineering hallway, a gift he knew would be useful simply because he'd taken those steps.

It could also be a thoughtful email you send, a helpful comment you leave, or a clever video you make—it doesn't have to be in-person or in real-time. Sincerely asking someone, "What's the most interesting thing you are working on right now?" or "What's your favorite and least favorite part of your job?" invites them to connect on their terms first. Treat your coworkers like customers (they are users of your ideas after all, at least that's what you're hoping) and border-walking as user research. Some of your walks will be ignored, or politely declined, but every time someone engages, and a connection is made, the feedback loops in the organizational system will reinforce you more than before.

PART 4

Make Design Easier

FRANK LLOYD WRIGHT, often called America's greatest architect, was tyrannically controlling of his projects. Even his wealthy clients, who hired him to design their mansions, had to endure restrictions over what furniture (also designed by Wright) they could have in their homes, and whether it could ever be rearranged. When Wright worked on the creation of the iconic Guggenheim Museum in New York City, he asked legendary artist Alexander Calder to make a giant mobile to hang from the central skylight. He met his match.

Wright demanded that the sculpture be made of gold. Calder rejected this idea because he always worked with industrial materials, and he thought something darker would look much better. So that's what he proposed. It was no surprise that Wright disagreed and reaffirmed his demand. Calder considered his response. At first he appeared to give in, but he had a clever twist: he agreed to make it out of gold, but only if he could paint it black. Wright, of course, refused. Years later, Calder would have the last word, exhibiting a massive sculpture at the Guggenheim Museum's central hall that was mockingly titled, *No! to Frank Lloyd Wright*.

We've refrained from stories about artists in this book until now for good reason. Art can be about the artist, but design is for the people. Wright and narcissists like him (he wore capes, you know) are only good examples of how to make design *harder*. A famous artist, whose clients hire him because of his fame, can get away with behaving poorly. The rest of us don't have that luxury, and even if we did, behaving this way doesn't lead to better work. Just ask Wright's clients, who sit in their beautiful living rooms in uncomfortable, angular wooden chairs that they can never move or replace. For us, it's a reminder that sometimes the ego trap is a childish wish for artistic entitlement, stuck in a grown-up's power trip, masquerading as a career. It would be wiser to make design the profession and art the hobby, where the art and the artist can be free.

Design Is an Infinite Game

In game theory, the term **finite game** means an activity where there is a clearly defined winner in a set period of time, like chess or checkers. Design is implicitly taught as a finite game. In every design project in school or at work, we want our ideas to "win": to be chosen or to solve the problem. When people reject our prototypes and sketches, we're disappointed and feel like we've lost. But there's a way of thinking of design as an infinite game, where the expectation is that there's no end.[1] Design as an infinite game transforms frustrations into opportunities, making design as a profession or an ambition much easier. If the game never ends, a "loss" becomes the biggest chance for growth you've had so far. Perfect outcomes teach us nothing: it's only when we struggle that we're forced out of our comfort zones.

A great example is the perennial frustration of answering the same basic questions about design from coworkers or friends throughout our lives. As a finite game, it feels repetitive and pointless, because even if we explain something well, we know we'll have to do it again with every new boss or person we meet. It's demoralizing and uninteresting and we dread it.

In an infinite game, there's never a final winner or loser. There's only time moving forward and things improving or declining. If you see teaching design as an infinite game, the same situation is a victory. The team you're leaving, the one it took years to influence into more design maturity, will be a delight for the designer who replaces you. That difficult VP, who you finally convinced to value aesthetics or usability, might start her own company and bring that perspective with her.

You might ask, "Why should I feel good about this kind of invisible work, helping some other team or project I may never see?" That's a question only someone playing a finite game would ask. You're forgetting all the situations you've been in where someone else did the heavy lifting before you got there. You are a beneficiary of the infinite game that all the people who came before you played. How do you think we get to the world W. E. B. Du Bois described,

---

1   James Carse, *Finite and Infinite Games* (Free Press, 1986).

where affluence, education, and health are in abundance? It's not by focusing on finite games.

Think about your answers to these questions:

- What inspired you to become a designer?
- Who created your job before you were hired?
- Who fought for that budget and defined the role?
- Who was the first designer at your organization?
- Who was the first designer with your gender or background?

All of our wisest teachers, Socrates, Hypatia, Buddha, Jesus, saw their teachings as infinite games. Noble pursuits never end because there's always someone who needs help or has lessons to learn. This means, of course, teaching design is endless. Every lost person you help find their way is one less burden for the next generation. Being an expert in any field that wants to be useful is playing an infinite game. Take joy in it. You can reframe every "failure" as an investigation you learn and grow from.

The Japanese have a concept called **wabi-sabi**, which is the beauty of accepting imperfections—it's only through failures that we develop the character we need to become better creators. Whoever your heroes are, we bet they inspire you because of their tireless commitment to their vision, not because they easily won a finite game years ago and continue to celebrate that single victory over and over again.

Reframing gravity problems as infinite games helps protect against burnout. You are not going to radically transform your company culture because of what does or doesn't happen in today's design critique. You're working within the organizational systems of dozens or hundreds of people, and you can only influence some of them. There will always be another project, another team, and another organization in your future, so don't be fatalistic about the challenges you're facing now. Focus on keeping your eyes open and making problems actionable: it's the wisest way to invest in the infinite game that is your career, your life, and your legacy. If you can, be more like Calder and less like Wright: have some fun and take yourself less seriously. Your work, and your life, will benefit from it.

Teach the Basics with Heart

Imagine if you had mind-control powers and could get your CEO, prime minister, mayor, landlord, or other powerful person with zero design awareness to permanently believe in any idea. What would you teach them? Would you choose to explain the difference between monospaced and display fonts? Or why user journey maps can be more valuable than ethnographic research methods? Of course not. You'd pick something simple because the basics have the most power. We get distracted by our expertise and the assumption that more "advanced" problems are the hard ones, but usually the opposite is true.

The basics, in the hands of the right person, transform organizations. That's what changes who the deciders are, the kinds of goals they choose, and who they hire. Remember that generalists have the power: they run your organization. Until they have a generalized understanding of good design and how it's done, progress around your interests is unlikely to happen.

Insecure designers often fear teaching people, because they believe if they help that cagey project manager learn basic UX skills, they won't be needed anymore. This is an attempt at career security through obfuscation, rather than value, which will eventually fail. It's also finite-game thinking. There will always be free websites and cheap online courses that falsely promise anyone can learn anything with just a few secret tricks (and five easy payments of $99.99).

Instead, put the Dunning-Kruger effect to work for you: once people learn basic knowledge, their overconfidence declines and respect grows. If we can assume the business model you are in benefits from higher-quality design, that project manager will recognize how important, time consuming, and difficult design work is and want to staff it properly. The infinite game leads to different assumptions, including the recognition that the best advocacy for design resources will come from generalists, not designers.

At conferences, a popular topic designers discuss is what frustrates us in our workplaces. This frames our experience as a finite game. But as an infinite game, the question becomes, "What is most valuable for powerful people to learn from us?" Framed this way, it's easier to stay motivated and to enjoy investigating how to do it well. If we can make design fun to learn, more people will seek this knowledge from us.

Here are three design lessons you will teach countless times, with our advice on how to make them more effective:

- **Design needs to be involved early.** First, take design out of this sentence. Think about the project leader's perspective—every role wants to be involved earlier! There is no one, not marketers, not engineers, saying, "Please tell me what's happening as late as possible so I can be surprised and ineffective." Instead, ask: "What goal does the project leader have that early design participation will achieve?" And how will your involvement not create a "too many cooks in the kitchen" problem? (Hint: facilitation.)
- **You are not the user.** Make a fast-paced highlight video showing actual customers struggling to use your product, particularly a core feature that leaders believe is easy to use. The raw emotional power of watching a paying customer

struggle will have more influence than usability reports or conversion funnel logs. Often stories have more power than data, especially when told by a persuasive storyteller, and a high-profile customer who is upset might just be the most persuasive storyteller there is.

- **What intuitive really means.** The word intuitive and the phrase "user friendly" are popular but empty. Teach leaders that intuitive means *familiar*: a product is only intuitive if its design is familiar to the person using it. For example, the cockpit of an Airbus 320 aircraft is intuitive to a trained pilot, but not to a teenager who has only played video games. It's the same cockpit design in both cases, but it's intuitive in only one scenario. The lesson for leaders is that if they want an intuitive product, half the battle is carefully identifying who they are designing for and what will be familiar to that audience. Nothing is intuitive to everyone.

Putting It All into Action: Situations and Tactics

Well, here we are in the home stretch of the book. Now that we've shared our big ideas and theories about why design is hard, we want to prepare you with some real-world scenarios. Based on our research and lived experience, these are the nine challenging situations you will face in your career. And as cranky design veterans who want to help you avoid pitfalls and primal scream therapy, we've provided tactical advice on how to handle each one. The secret is to reframe every situation as a creative problem-solving challenge, and to remember that you are good at creative thinking.

No one knows what you do or what value you add

What happened. Executives sometimes do stupid things, like hiring people without fully understanding why. They might be copying what competitors are doing or get inspired by a trend they heard on their favorite podcast. This can result in a situation where, despite good intentions, no one on the team has the expertise to properly onboard or lead a new designer. Alternatively, the CEO could be paying for design theater: they want to be able to say to customers, "we have a UX team," or "our usability experts approved this," without actually having to change anything about how product decisions are made. This explains why it's common to join a team, yet no one has any idea why you're there or what you're supposed to do. Everyone assumes someone else has done the thinking about how to set you up to succeed, and the Dunning-Kruger effect means well-intentioned people don't realize what they don't know.

Why this is hard. You may feel like you've been tricked. You will certainly feel alone and unsupported, and that you're forced to play finite games to survive, which is scary.

What not to do. Don't rush to give a presentation to raise awareness of design. This is not a good first move because it can perpetuate the stereotype of design's trivialness and ego-centrism. What people do like is having their problems solved. Until you've solved a problem for someone, even if it's a small one, you're not ready to ask for more attention from the team. Live your case study before you present it: their problems are not in Figma, so get onto their turf to find them.

Find a partner. When you join a project, interview every contributor like they are a customer. Do border-walks first. In Leah Buley's book, *The User Experience Team of One: A Research and Design*

Survival Guide, she calls this a listening tour.[2] Ask them to show you, not tell you, what they make. By having them show you their work and how it appears to users, you will be in a visual medium, a strength for you. Be curious, take notes. Ask where they think they need design help. Who is friendly to you? Who welcomes your input? What is the smallest unit of success you can create? When someone makes a unit of work better because of you, that's what you want to present about. When your team talks about your impact in positive terms when you're not around, the landscape is changing in your favor.

What to do next time. Bet big on your investigative powers. Before you join a team, talk to the project leader about the role they expect designers to play. Ask them about good and bad experiences with past designers. Work on a plan for how you and your abilities can directly help their goals.

You're told there isn't time to do good UX work

What happened. You asked to do much needed research or design explorations, but the project leader told you no because there wasn't time.

Why this is hard. Who wants to be forced to do bad work? It's demoralizing.

What not to do. Don't act surprised or offended. The job of being a project leader involves saying no far more often than yes. There could be very good reasons why resources need to go elsewhere. Designers hate to have their expertise doubted, but we often do it to other roles.

---

2   Leah Buley, *The User Experience Team of One: A Research and Design Survival Guide* (Brooklyn: Rosenfeld Media, 2013).

What to do next time. Is it possible everyone feels this way? Engineers also want to build high-quality products. Maybe they're upset too and can become allies. Realize that "there isn't time" is a poor excuse that leaders use to end conversations. The better question is, "What is important enough that it gets time and how is that prioritized?" There's always enough time to do something, just not everything. Using *their* language, describe the value of your proposal in terms of the project goals. If you're still refused, ask how to raise this issue earlier so it can be considered as an explicit goal on the next project.

Decisions are made without you

What happened. Well, you weren't there. So you don't really know, do you?

Why this is hard. Who wants to be left out?

What not to do. Don't take it personally. Avoid making an assumption that people think you'd cause friction by attending, or that you were excluded because someone doesn't like you.

Why it happened. One simple reason may be that good leaders want as few people in the room when hard decisions are being made. They will pick people who are most helpful in considering multiple aspects of a decision (which are generalists, or people good at thinking beyond their own specialization). The concept "too many cooks in the kitchen" is a real problem leaders try to avoid.

What to do next time. Put your ego aside and ask for feedback about what value you'd need to add to be included. Explain how being excluded hurts the project's goals (is this really true?). Maybe there's something you don't know. Is there a gap between how useful you think you are compared to how others perceive you? Was it clear to the others that you expected to be involved? You won't know unless you ask.

Use your allies. Rely on your skills for influence and negotiation. Who would benefit most from having you there, and how can you convince them it's in their best interest to include you? Or consider: do you really need to be in the room when the decision is made? Wouldn't it be better if someone more powerful, who you trusted, was there to represent your interests? Sometimes you benefit from the decision being made without you. If you can, reframe the issue so it can be solved. Often, "big meetings" are just theater, and the real decision was made in some back room the day before. Who can represent you wherever the decision is actually made?

Your team is great, but you're struggling to find good ideas

What happened. You found a great place to work, but you're experiencing a creative block.

Why this is hard. Maybe because there's nowhere to hide or no one to blame? When the team has obvious problems it's easy to put your attention on how it's preventing you from doing good work.

What not to do. Don't think of yourself as a fraud. **Imposter syndrome**, a feeling of anxiety that your skills aren't good enough, is rampant in creative fields. We have unrealistic expectations that creative work will be a constant state of flow and inspiration, but the truth is that feeling stuck is part of the job for all people who work with ideas. No one is immune. As artist Chuck Close said, "Inspiration is for amateurs. The rest of us just show up and get to work."

Be a designer, not an artist. Your job is to improve things, which rarely requires inventing things no one has made before, or having a grand vision that radically improves everything. If you're stuck, spend more time watching, not just listening to, customers use your product. What obvious things are broken or confusing? There is usually plenty of low-hanging, tasty fruit if you look in the right place. Ask customer support what issues take up most of their time. Or go study your competitors' products: what do they do better that you can learn from?

No one listens to your suggestions

What happened. Let's be realistic. If no one ever took your suggestions and you provided no value, you wouldn't have a paycheck. You're just feeling ignored and you don't like it.

Why this is hard. Rejection is painful. It can feel like you don't exist or your thoughts don't matter.

What not to do. Don't assume it's about you. Did others get their suggestions rejected too? Maybe you work for someone who defends the status quo as if their life depends on it. All decisions are about tradeoffs, and business, engineering, and leadership can all have valid but conflicting opinions. Unless your suggestions incorporate multiple perspectives, you're leaving it to the decision-maker to balance tradeoffs across the different points of view each job role tends to have.

Investigate. Did you make suggestions to the right person? At the right time? Were they in the right mood? A good place to start is to consider that perhaps you don't understand the problem they're trying to solve. What is the user experience for the person listening to your suggestion?

What are their mental models and goals? What are their frustrations? Use your curiosity and creative skills to figure out the right place, time, medium, and style to offer your suggestions so they're most likely to be effective.

What to do next time. What is your relationship with the decision-maker? How much do they trust you? What past experience do they have taking your suggestions and having success? Decisions are as much about the relationship as the idea itself. Reach out and ask for feedback on how your suggestions can be more useful next time, instead of focusing on the limitations of your coworkers.

Someone challenges you on a design/UX principle

What happened. In a design review for MegaStink8, the powerful engineering manager (who apparently hates your customers) makes a crazy request to advertise the latest new features. He asks you to design a pop-up window with a sliding 3D-animated carousel, which plays for five minutes, while graying out the rest of the screen so users are forced to watch it until they hit the close button. He's seen this in other products and wants it included, even though it doesn't fit within the overall design or good usability guidelines.

Why this is hard. It can feel disrespectful and demoralizing. You have design expertise you were hired to use, but you're being ignored (in Part 2, Jesse had a similar situation).

What not to do. Don't make it a binary choice between making the change without a conversation or berating your engineering partners on how they don't understand UX. There's an opportunity to approach this as a different problem to solve while maintaining relationships and leveraging your skills.

Insight. It's understandable for people to see design choices in other successful products and assume all those choices are good ones, or are applicable in your case. The fact that they have an opinion reflects their curiosity about design. They just need someone with expertise to guide them: this is your golden opportunity to teach the basics with heart (think infinite game).

What to do. Ask what problem they are trying to solve. Point out that this choice will likely create other problems that are worse. If it's bad for usability, it's bad for your bottom line. If you have core UX tasks that are considered the most important to customers, refer back to them and see how "discovering new features" fits in. Alternatively, is there an ally in the room who you know will argue your side? If so, invite their opinion first. Maybe you don't need to fight this battle yourself.

They think they can do your job because they've learned a tool

What happened. Without talking to you, your product manager fired up Figma, did some mock-ups, and sent them to engineering to whip up an A/B test for a quick feature change.

Why this is hard. It can feel like betrayal: someone is making you look bad or ignoring your expertise.

What not to do. Don't jump to conclusions about why they did it. From their perspective, they're trying to be productive and help the project. They most likely do not understand why this might offend you. Engineers are often not territorial about code, so that mindset can spread into other roles.

Insight. It's probably a mistake to define yourself based on a tool or a process since they eventually get replaced: it's a self-limiting reputation. Instead, it's preferable to be seen as someone who makes good decisions, has good advice, and gets things done with whatever tool needs to be used. Make your value extend beyond using software.

What to do next time. Check in on the alignment of their goals and your goals. Walk them through the steps you would have taken to do this task, and why the quality would have been higher had you been involved. Is that quality valued in this case? Do a design review with the product manager. Set the expectation that design ideas go through lightweight reviews (just like code reviews). Share that product changes need to fit inside the whole strategy. Will this change hurt things further down the conversion funnel? Does it support or violate UX principles that were agreed on for the product?

You got your seat at the table but discover the people there are fools

What happened. You've worked hard for months to get into the meetings where decisions are made. Finally, you arrive, and you're shocked! People are behaving poorly. There are power struggles and petty disputes. You thought you'd won, but now you're starting to see how the organization really works—or, rather, doesn't work—and you're disappointed.

Why this is hard. Your faith in leaders has been broken. It can make you question other assumptions about your workplace.

What not to do. Don't give up. This feeling is normal: you've just leveled up in the video game of being an impactful designer, so the challenges have gotten harder. This is a good problem!

Insight. As the amount of power in the room rises, often the amount of dysfunction rises too. The Peter principle, where people are promoted to their level of incompetence, is real. You may encounter more people obsessed with their careers, who sacrifice healthy, balanced lifestyles for their work ego. Consider this: who is going to fight to protect their workplace power as if their life depends on it? The people who have no other life.

What to do. Good managers shield their teams from executive chaos and make it seem like things are sane. Look for them. Seek out allies and look for people who share your sensibilities. At this table, you have better opportunities because you have closer contact with more power and influence. You can now make more accurate maps and discover better leverage points.

The resources you were promised disappear

What happened. You toiled for months to line up stakeholders behind your big strategy. They promised you power, influence, budget, and staff, too. Then, suddenly, it all falls apart.

Why this is hard. It can feel like you've wasted significant time. It seems like your partners are on a different planet, and your company will never become a real design-led organization that takes its customers seriously.

What not to do. Don't assume you were betrayed, or that the person who made the promises is the one to blame. They may still be the best ally you have, they just didn't win this time. It's likely that they lost something too, possibly something bigger than what you lost.

Insight. There are much higher priorities in your organization than what *you* think is the highest priority. Executives may not like each other, senior leaders may change their minds, or a powerful customer reverses their commitments. Your plan may have been rejected for reasons that have nothing to do with it, even though it impacts you significantly. It can be hard to explain this to the people you made promises to, but sometimes them's the breaks.

Negotiate. If a deal falls apart, guess what? There are new deals to be made! If you act fast enough, maybe you can get part of what you need from a different source.

What to do next time. If you're surprised by what happened, it means there's something you didn't see. Learn where you were out of sync with the key players involved. What problems are they trying to solve that you've not yet understood or taken seriously? Use your curiosity to find out.

PART 5

Design in the Real World

HERE, IN THE fifth and last part, we have one final confession. The truth is that every word, idea, and image in this book was generated by a new hybrid GPT/Llama9 model LLM AI system, known as StinkBrain (made by MegaStinkCo). The actual "authors" of this book were barely involved and are currently on a secluded beach in Hawaii, lying in the shade, fruity umbrella drink in hand, happily not thinking about design or anything else.

Hopefully it's obvious, to our beachless disappointment, that this is a joke, but we do actually have something important to disclose. There is a limitation in all books, which is that we, the authors, can't address your unique reality. It's easier to learn lessons in the safety of books, where it all seems to make good sense, than to apply them to the chaos of real life. What good is "excellent advice" if it turns out not to apply to your reality, or it makes things worse? Not very good at all.

Alfred Korzybski, the inventor of the field of semantics, famously wrote that "the map is not the territory." It's all too easy to fall in love with our maps and forget that all maps have flaws. They are distortions of the truth, carefully chosen, we hope, to make important things easier to understand, but that always comes with tradeoffs. Books can be good references for when you feel lost, but we don't recommend that you assume we're right about everything. We'd much rather you think it through and come to your own conclusions, even if they disagree with ours.

At the start of this book, we presented you with our core premise: design is hard and there are only three choices for what to do about it. Now that we're at the end of our time together, here they are again for review:

A. Seek power. Being undervalued means you do not have the power you need. Who makes decisions you think you should be making? Who doesn't listen to your advice but really should? Designers need power to design. There's no way around it. Decisions are really about power, and you need to increase how much power you have.

B. Become influential. If you don't want the responsibilities that come with power, that's OK. Instead, become an influencer. Think of your job as a consultant or an advisor, and draw from the rich heritage of skills those roles have always had. If the powerful people you work with listened to you 30% or 50% more often, and gave you more credit, would you enjoy your career more? If yes, then influence is the way.

C. **Be self-aware.** Even without wanting more power (A) or influence (B), if you can mature your beliefs about design and escape the ego trap, you'll become a healthier person. Your career will have more flow and be more fulfilling. You'll get smarter at identifying healthy places to work, or perhaps you'll realize you want to be your own boss. By becoming self-aware, you'll be less reactive to the messy reality of human nature in organizations.

Early in the book it may have seemed like these choices were mutually exclusive, but ultimately, this trio works together. Whether you rise as a generalist in your organization or become the best specialist you can be, you need to know who you are, and what you really want, to gain power and to influence effectively.

Maybe you know in your heart that you just want to be the best craftsperson, or you realize your ideal environment is a small startup where influence is easier. Perhaps instead you're ready to lead and want to rise in the ranks of your Fortune 500 company or local government. This is good news, as society needs people who are excellent at their craft, who work well with others, and who want to make a better world. Regardless of the path you choose, you now have a better understanding of how systems work, and tools and strategies to navigate your role.

Cash Money, Baby!

We've put it off as long as we could, but any serious discussion about design as a job should include compensation. You may have been asking all along, "Wait a second, why do all this work to influence people if I won't be paid more for it?" Who deserves to get paid what amount and why is a lightning rod, which is why most design books pretend we live in some utopian world where money no longer exists.

Not getting paid what you feel you deserve can make design harder, so here is our opinion.

First, we think much of the advice in this book doesn't take much extra effort to follow, since it's more about an attitude than exertion. The beauty of reframing is that the same effort, shaped differently, can yield far better results. And that extra effort might decline once you get the hang of it (and you're valued for the outcomes you produce, rather than just your output). Having more power and influence might make you more fulfilled and effective at work, even without more pay. And if you're more valued and recognized by deciders as a result, raises and promotions might follow on their own.

Second, in most organizations, everyone feels underpaid! Even some brazen CEOs of S&P 500 corporations, who have average salaries that are staggeringly 272 times greater than their employees, feel they deserve to be paid more.[1] Of all the things unique to you, how your organization and your boss decide to compensate employees is high on that list. Pay is ultimately something you have to negotiate for, and we can say, as discussed in Part 2, that it is hard to negotiate based on *should* arguments. Why? Everyone feels they should be paid more, and no one is volunteering to be paid less.

Given the perhaps unfair reputation the design profession has, this may be a chicken-and-egg problem to overcome. Before you're paid more, you may have to demonstrate that you're providing more value than a run-of-the-mill, ego-trapped, output-worshipping,

---

1 Jena McGregor, "CEOs Who Feel Underpaid Are More Likely to Lay Off Workers," *Washington Post,* August 23, 2018, https://www.washingtonpost.com/business/2018/08/23/ceos-who-feel-underpaid-are-more-likely-lay-off-workers/; Sarah Anderson and Alan Barber, "The CEO Pay Problem and What We Can Do About It," Congressional Progressive Caucus Center, https://www.progressivecaucuscenter.org/the-ceo-pay-problem-and-what-we-can-do-about-it/#big.

gravity-problem-denying, elitist, overly specialized, system-avoidant, detail-obsessive, standoffish, factionalizing designer. That may be the only kind of designer your organization has met or even heard of before you arrived. This is an investment you have to decide you are willing to make. If you are, it helps to have allies in your organization who feel they benefit from your skills, as then they'll have a vested interest in helping you succeed.

Study Your Landscape

To dig deeper into maps and territories, we can learn from professions who must close the gap between them regularly. Journalists, anthropologists, and even Navy SEALs are trained to first study the landscape they're on before using any of their primary skills. Why? The landscape always wins. The landscape is bigger and stronger than any person or ability. If you're prepared for survival in the desert, but you don't notice you've landed in Antarctica, you can work very hard using all your expertise and still fail. If you work against your landscape, or are ignorant of it, you will lose.

In Part 3, we said "systems come first," and that might be the best place for you to start. Study the people and culture you're in to see how they really work, and how what they say differs from what they do. Save your energy for when you know where to invest it to have the highest chances of positive results (i.e., seek leverage points). Do what you're asked to do at first until you see what angle is best to play.

Even if you've been in the same job or organization for awhile, it's never too late to do this. You're a new person in each moment, and you're only bound to your past identity if you choose to be. You can decide that today is the first day of the rest of your life, and you're now the wisest version of yourself you've ever been.

Take a fresh look around and ask:

- Who is effective at making things happen here?
- What do they do that others don't?
- Where is the interesting friction? Like Dieter Rams, who should I buy brandy for?
- What metaphors and language are popular? How can I speak like a local?

And about yourself:

- What makes me feel fulfilled?
- How do my values align with my work?
- What are my real strengths and weaknesses?
- How am I building trusted relationships?
- How can I express my abilities in ways that people value?
- What community is helping me grow?
- How can I give something back and be a better ancestor?

The Last Part of The Last Part

Here at the end of our book, we hope you now understand why design is hard and see ways to make it easier. We don't think it ever gets *easy* though. Most of the important challenges in life aren't easy either and that's part of why we do them. Wisdom might just mean that we go about these pursuits in a healthy way, with open hearts and minds.

If this book helped you, please tell others about it. And as you find your way, share your story with us so we can learn from you. Join us at designishard.com.

Recommended Reading

***How Design Makes the World*, by some guy named Scott Berkun (Berkun Media, 2020).**

The best book for explaining good design to anyone. A short and fun successor to books like Don Norman's *Design of Everyday Things* and Steve Krug's *Don't Make Me Think*. Great for teams, clients, executives, and friends. They'll learn the big lessons about design you wish they knew. A great companion to *Why Design Is Hard*.

***The User Experience Team of One: A Research and Design Survival Guide*, by Leah Buley (Rosenfeld Media, 2013).**

The advice in this book is grounded in the real world, something rare for design books. Anyone who works in UX should read this book, even if they work on a larger UX team. It exemplifies how to escape the ego trap and reframe frustrations into solvable problems.

Extra Bold: A Feminist, Inclusive, Anti-Racist, Nonbinary Field Guide for Graphic Designers, **by Ellen Lupton, Jennifer Tobias, et al. (Princeton Architectural Press, 2021).**

> This eye-opening combination of design history and practical advice will change how you think about your profession. It's recommended reading for all designers, graphic or otherwise. It will widen your sense of what design means, of who has contributed to the profession, and who we need to do it in the future.

Communicating the UX Vision: 13 Anti-Patterns That Block Good Ideas, **by Martina Hodges-Schell and James O'Brien (O'Reilly Media, 2013).**

> A practical, advice-filled book focusing on how poor communication is the real challenge most designers have to overcome. The anti-patterns they define are a checklist of situations that limit many designers' careers, and their excellent, grounded advice will help you overcome them.

Articulating Design Decisions: Communicate with Stakeholders, Keep Your Sanity, and Deliver the Best User Experience, **Tom Greever (O'Reilly Media, 2015).**

> With wisdom and insight, Greever explains how to handle many of the communication challenges you'll experience, including dealing with stakeholders, navigating design meetings, and improving listening skills.

***Orbiting the Giant Hairball: A Corporate Fool's Guide to Surviving with Grace*, by Gordon MacKenzie (Viking, 1998).**

If you feel the corporate world is dragging you down and destroying your creativity, this is the book for you. Written by a former executive at Hallmark, he explains how he used the metaphor of a giant hairball to help him accept and work with the limitations of large organizations. It's only in print/physical editions, but it's worth it for his hand-drawn art on many of the pages.

***Getting to Yes: Negotiating Agreement Without Giving In*, by Roger Fisher, William L. Ury, and Bruce Patton (Penguin, 2011).**

This classic guide to negotiation will help you recognize how much of your work is negotiating with other people, and improve your ability to find mutually beneficial ways to get your needs met.

***Art & Fear: Observations on the Perils (and Rewards) of Artmaking*, by David Bayles and Ted Orland (Image Continuum Press, 2001).**

This short book was a primary conceptual model for *Why Design Is Hard*, in that it is a kind, generous, and honest book written by a pair of authors. The book itself is useful for designers in exploring their emotions about their craft, as we recommended in Part 2. However, good design, unlike art, serves the needs of the customer and the client first and the creator second, so read it with that critical constraint in mind.

***How Designers Think*, by Bryan Lawson (Routledge, 2005).**

As the title suggests, this is a deep inquiry into the habits and patterns of behavior that go on when designers are doing their work. It's a great reference on both problem solving and better understanding the psychology of design work.

Acknowledgments

Thanks to the folks who kindly let us interview and/or torture them with early drafts (Design Hero Manifesto RIP): Anne Hjortshoj, Lisa deBettencourt, Bob Baxley, Dana Chisnell, Paolo Mayubolo, Consuelo Valdez, Billie Mandell, Jenna Hammer, Lisa Angela, Tanya Johnson, Veronica De la Peña, Vanessa Riley Thurman, Donna Givens, and Mike Knecht.

Thanks to Marlowe "escalope" "derp" Shaeffer, who went above and beyond: she's the superhero of editors. Cheers to Tash Willcox for the cover design and diagrams. Thanks to Zoe Norvell for the interior design. And gratitude to Amy Berkun, Alyssa Fox, Kathy Gill, Whitney Hess, George Perantatos, Adrian Klein, and Nicky Bleiel for proofreading. Thanks to Terri Rothman for help with image rights and permissions.

Cheers to everyone who joined our Substack (https://whydesignishard.substack.com/) and gave us feedback along the way.

Scott Berkun would like to thank Jill Stutzman, Judi Berkun, Todd Berkun, Chris McGee, Vanessa Longacre, Marlowe Shaeffer, Neil Enns, Chad Urso McDaniel, Bryan Zug, Shawn Murphy, Zac Cohn, John Lykins and Christiana Birkeland for helping him restore his sanity from 2021 to 2022. He wasn't sure he'd ever want or be able to write again and this book is a testament to his recovery.

Bryan Zug would like to thank his family (Jen, Rue, and Thomas), along with all of the coworkers and clients he's been in the trenches with since building his first web business in the small town of Trinidad, Colorado in 1995.

About the Authors

Scott Berkun is a former project leader and UX designer. He is the bestselling author of eight books, including *How Design Makes the World*, *The Myths of Innovation*, *Confessions of a Public Speaker*, and *The Year Without Pants: WordPress.com and the Future of Work*. His work has appeared in The New York Times, The Washington Post, Forbes, The Wall Street Journal, The Economist, The Guardian, Wired Magazine, USA Today, Fast Company, National Public Radio, CNN, NPR, MSNBC and other media. His popular blog and email list are at *scottberkun.com*.

Bryan Zug is an industry veteran in product development, UX design, and leading generous and high-performing teams. He cofounded Zillow's first DesignOps team and helped standardize product development practices across the company. He was a senior UX Program Manager at Amazon.com, leading DesignOps for creating great software and experiences for Fulfillment Center staff around the world.

He founded his own video and livestreaming production company, Bootstrapper Studios, before working at Startup Weekend / Techstars as director of customer experience. He is also the cofounder of Ignite, the short format presentation event that takes place in hundreds of cities around the world.

Image, Photo and Design Credits

All illustrations by Tash Willcocks, based on sketches by Scott Berkun

Cover design by Tash Willcocks, based on a concept by Scott Berkun

Citibank sketch, Paula Scher/Pentagram, used by permission

Citibank logo, Pentagram/Citigroup Inc.

Dieter Rams photo, used by permission from ©vitsoe - www.vitsoe.com

Interior fonts: The body text of this book is set in Source Sans Pro, designed by Paul D. Hunt for Adobe. Chapter titles and headers are set in Halyard Display, designed by Joshua Darden of Darden Studio.

Made in the USA
Las Vegas, NV
01 October 2024